Naturally Speaking

A Dictionary of Quotations
on
Biology, Botany, Nature and Zoology

About the Compilers

Carl C Gaither was born in 1944 in San Antonio, Texas. He has conducted research work for the Texas Department of Corrections and for the Louisiana Department of Corrections. Additionally he worked for ten years as an Operations Research Analyst. He received his undergraduate degree (Psychology) from the University of Hawaii and has graduate degrees from McNeese State University (Psychology), North East Louisiana University (Criminal Justice), and the University of Southwestern Louisiana (Mathematical Statistics).

Alma E Cavazos-Gaither was born in 1955 in San Juan, Texas. She has worked in quality control, material control, and as a bilingual data collector. She is a Petty Officer First Class in the United States Navy Reserve. She received her associate degree (Telecommunications) from Central Texas College and her BA (Spanish) with a minor in Art from Mary Hardin-Baylor University.

Together they selected and arranged quotations for the books *Statistically Speaking: A Dictionary of Quotations* (Institute of Physics Publishing, 1996), *Physically Speaking: A Dictionary of Quotations on Physics and Astronomy* (Institute of Physics Publishing, 1997), *Mathematically Speaking: A Dictionary of Quotations* (Institute of Physics Publishing, 1998), *Practically Speaking: A Dictionary of Quotations on Engineering, Technology, and Architecture* (Institute of Physics Publishing, 1998), *Medically Speaking: A Dictionary of Quotations on Dentistry, Medicine and Nursing* (Institute of Physics Publishing, 1999) and *Scientifically Speaking: A Dictionary of Quotations* (Institute of Physics Publishing, 2000).

About the Illustrator

Andrew Slocombe was born in Bristol in 1955. He spent four years of his life at Art College where he attained his Honours Degree (Graphic Design). Since then he has tried to see the funny side to everything and considers that seeing the funny side to science has tested him to the full! He would like to thank Carl and Alma for the challenge!

Naturally Speaking
A Dictionary of Quotations
on
Biology, Botany, Nature and Zoology

Selected and Arranged by

Carl C Gaither
and
Alma E Cavazos-Gaither

Illustrated by Andrew Slocombe

Institute of Physics Publishing
Bristol and Philadelphia

IOP Publishing Ltd has attempted to trace the copyright holders of all the quotations reproduced in this publication and apologizes to copyright holders if permission to publish in this form has not been obtained.

British Library Cataloguing-in-Publication Data
A catalogue record for this book is available from the British Library.

ISBN 0 7503 0681 5

Library of Congress Cataloging-in-Publication Data are available

Commissioning Editor: James Revill
Production Editor: Simon Laurenson
Production Control: Sarah Plenty
Cover Design: Frédérique Swist
Marketing Executive: Colin Fenton

Published by Institute of Physics Publishing, wholly owned by The Institute of Physics, London

Institute of Physics Publishing, Dirac House, Temple Back, Bristol BS1 6BE, UK
US Office: Institute of Physics Publishing, Suite 1035, The Public Ledger Building, 150 South Independence Mall West, Philadelphia, PA 19106, USA

Typeset in TeX using the IOP Bookmaker Macros
Printed in Great Britain by J W Arrowsmith Ltd, Bristol

I dedicate this book to my long time friend Diana Kiaha-Watts

Carl C Gaither

I dedicate this book to my long time friend Annette Kutz

Alma E Cavazos-Gaither

To the nature lover the universe constantly pours out its wealth. Daily he gathers the fruits of seed sown in the beginning of the world.

For him no season is dull, for each is successively absorbing: In Spring he is entranced by the awakening of myriad forms of life; Summer reveals the maturity of all creation, Autumn brings the fulfillment of earlier promises; Winter lulls life to sleep, with its assurance of the resurrection.

All weathers are one: The rains of Spring nourish all nature; the heats of Summer mature and ripen its fruits; the frosts of Winter give rest and peace; in all he rejoices. Each day is good: In the morning life awakens with him; through the noon it works; the peace and quiet of evening shed their benediction upon him.

He knows no dull moments; he seeks not to hurry time. If he be delayed, he may discover something never before seen by man, and his impatience is forgotten. His youth is filled with the joys of discovery; in middle age the marvels about him hold his interest undimmed; he awaits old age with calmness, for he is one with the universe, and is content.

Preble, Edward A.
Nature Magazine
The Lover of Nature (p. 537)
Volume 50, Number 10, December 1957

CONTENTS

PREFACE

History is replete with anecdotes and bons mots relating to statesmen, soldiers, artists, philosophers, and more other types of notables; but even a well-informed man finds it difficult to enliven talk with quotations from scientists.

Dubos, René
The Dreams of Reason
Chapter 3 (p. 40)

Naturally Speaking: A Dictionary of Quotations is the largest compilation of nature quotations published to date. The purpose of this book is to present quotations so that the reader can gain an idea as to the depth, width and breadth of the subject of nature. Additionally the book attempts to provide answers to the questions of "Who said what?" and "Where does it come from?".

There are many books of quotations and a large number of them have but a small section on nature and nature-related topics. *Naturally Speaking* is a quotation book that is devoted completely to the topic of nature. These quotations are gathered from many sources and were chosen because of the thought expressed or because the quotation expressed a truth that is generally recognized and approved. Hence, the quotations in this book can become a time-saver for the reader. By the use of this book, the reader may find a single terse phrase which expresses an idea or opinion that would otherwise need to be stated in several sentences of original composition. For the reader who is reading the book for enjoyment, *Naturally Speaking* becomes a master of ceremonies by introducing the reader to the great number of thoughts that lie within the pages of the book.

With so many well-written books of quotations on the market is another book of quotations necessary? We and our publisher agreed that there was a need since the standard dictionaries of quotations, for whatever cause, are sorely weak in providing entries devoted to quotations on nature. *Naturally Speaking* fills that need.

The understanding of the history, the accomplishments and failures, and the meanings of nature requires a knowledge of what has been said by the authoritative and the not so authoritative philosophers, novelists, playwrights, poets, scientists and laymen about nature. Because of the multidisciplinary interrelationships that exist it is virtually impossible for an individual to keep abreast of the literature outside of their own particular specialization. With this in mind, *Naturally Speaking* assumes a particularly important role as a guide to what has been said in the past through the present about nature.

Naturally Speaking is not only confined to the student or practitioner of Natural History but was designed also as an aid for the general reader who has an interest in nature topics. The general reader with no knowledge of nature who reads *Naturally Speaking* can form a pretty accurate picture of what nature is. Students can use the book to increase their understanding of the complexity and richness that exists within the scientific disciplines. Finally, the experienced scientist will find *Naturally Speaking* useful as a source of quotes for use in the classroom, in papers and in presentations. We have striven to compile the book so that any reader can easily and quickly access the wit and wisdom that exists and a quick glance through the table of contents will show the variety of topics discussed.

A book of quotations, even as restricted in scope as *Naturally Speaking*, can never be complete. Many quotations worthy of entry have, no doubt, been omitted because were unaware of them. However, we have tried to make if fairly comprehensive and have searched far and wide for the material.

Quite a few of the quotations have been used frequently and will be recognized while others have probably not been used before. All of the quotations in *Naturally Speaking* were included with the hope that they will be found useful. The authority for each quotations has been given with the fullest possible information that we could find so as to help you pinpoint the quotation in its appropriate context or discover more quotations in the original source. When the original source could not be located we indicated where we found the quote. Sometimes, however, we only had the quote and not the source. When this happened we listed the source as unknown and included the quotation anyway so that it would not become lost in time.

How to Use This Book
1. A quotation for a given subject may be found by looking for that subject in the alphabetical arrangement of the book itself. This arrangement will be approved, we believe, by the reader as making it easier to locate a quotation. To illustrate, if a quotation on 'botanist' is wanted, you will find six quotations listed under the heading 'botanist'. The arrangement of quotations in this book under each subject heading

constitutes a collective composition that incorporates the sayings of a range of people.

2. We were certain that good indexing was going to be an important aspect of our books. We recognize that without good indexing a book of quotations is nothing but a labyrinth without guidance where to find the material that is being sought. Hence, we have provided two indices (a) a SUBJECT BY AUTHOR INDEX and (b) an AUTHOR BY SUBJECT INDEX.

3. To find all the quotations pertaining to a subject and the individuals quoted use the SUBJECT BY AUTHOR INDEX. This index will help guide you to the specific statement that is sought. A brief extract from each quotation is included in this index.

4. It will be admitted that at times there are obvious conveniences in an index under author names. If you recall the name appearing in the attribution or if you wish to read all of an individual author's contributions that are included in this book then you will want to use the AUTHOR BY SUBJECT INDEX. Here the authors are listed alphabetically along with their quotations. The birth and death dates are provided for the authors whenever we could determine them.

Thanks

It is never superfluous to say thanks where thanks are due. Firstly, we want to thank Jim Revill of IOP Publishing who has assisted us so very much with our books. Next, a most substantial debt of gratitude is extended to the following libraries for allowing us to use their resources: The Jesse H. Jones Library and the Moody Memorial Library, Baylor University; the main library of the University of Mary-Hardin Baylor; the main library of the Central Texas College; the Perry-Castañeda Library, the Undergraduate Library, the Engineering Library, the Law Library, the Physics–Math–Astronomy Library, and the Humanities Research Center all of the University of Texas at Austin.

Samuel Johnson has written:

The greatest part of a writer's time is spent in reading, in order to write: a man will turn over half a library to make one book.

In James Bosewell
Life of Johnson
Volume I
6 April 1775 (p. 581)

We are sure that Joe Gonzalez, Matt Pomeroy, Chris Braun, Ken McFarland, Craig McDonald, Kathryn Kenefik, Brian Camp, Robert Clontz, and Gabriel Alvarado of the Perry-Castañeda Library certainly must believe that Johnson's statement is true since they had to put up with us when we were checking out the hundreds of books. Finally, we wish to thank our children Maritza, Maurice and Marlynn for their assistance in finding the books we needed when we were at the libraries.

A great amount of work goes into the preparation of any book. When the book is finished there is then time for the editors and authors to enjoy what they have written. It is hoped that this book will stimulate your imagination and interests in matters about nature and this hope has been eloquently expressed by Helen Hill:

If what we have within our book
Can to the reader pleasure lend,
We have accomplished what we wished,
Our means have gained our end.

In Llewellyn Nathaniel Edwards
A Record of History and Evolution of Early American Bridges (p. xii)

In closing we wish to leave you with these words from Jerry Flack.

Let us give students ideas worthy of their contemplation. The erudition found in quotations down through the ages is grist for the thinking of today's youth.
Teaching K-8
Quotations in the Classroom (p. 60)
Volume 24, Number 3, November/December 1993

Carl Gaither
Alma Cavazos-Gaither
October 30, 2000

ACCURACY

Carroll, Lewis
"How is bread made?"

"I know *that*" Alice cried eagerly. "You take some flour—:

"Where do you pick the flower?" the White Queen asked. "In a garden, or in the hedges?"

"Well, it isn't *picked* at all", Alice explained; "it's *ground—*"

"How many acres of ground?" said the White Queen. "You mustn't leave out so many things."

The Complete Works of Lewis Carroll
Through the Looking-Glass
Chapter IX (p. 254)

Darwin, Charles
...I value praise for accurate observation far higher than for any other quality...

In Francis Darwin (ed.)
The Life and Letters of Charles Darwin
Volume II
Darwin to Hooker
December 11, 1860 (p. 148)

...good heavens, how difficult accuracy is!

In Francis Darwin (ed.)
The Life and Letters of Charles Darwin
Volume II
Darwin to Gray
June 3, 1874 (p. 457)

Accuracy is the soul of Natural History. It is hard to become accurate; he who modifies a hair's breadth will never be accurate... Absolute accuracy is the hardest merit to attain, and the highest merit.

In Francis Darwin (ed.)
More Letters of Charles Darwin
Volume II
Darwin to Scott
November 26, 1868 (p. 323)

Gombrich, E. H.
Everyone is acquainted with dogs and horses, since they are seen daily. To reproduce their likeness is very difficult. On the other hand, since demons and spiritual beings have no definite form, and no one has ever seen them, they are easy to execute.

Art and Illusion
Part II, Chapter VIII (p. 269)

Hume, David
Accuracy is, in every case, advantageous to beauty, and just reasoning to delicate sentiment. In vain would we exalt one by depreciating the other.

An Enquiry Concerning Human Understanding
Section I, 8 (p. 90)

Huxley, Thomas H.
Accuracy is the foundation of everything else. . .

Collected Essays
Volume III
Science and Education
Technical Education (p. 432)

Johnson, Samuel
He who has not made the experiment, or who is not accustomed to require rigorous accuracy from himself, will scarcely believe how much a few hours take from certainty of knowledge, and distinctness of imagery; how the succession of objects will be broken, how separate parts will be confused, and how many particular features and discriminations will be compressed and conglobated into one gross and general idea.

A Journey to the Western Islands of Scotland (pp. 239–40)

Smith, Theobald
. . . it is the care we bestow on apparently trifling, unattractive and very troublesome minutiae which determines the result.

In W. Bulloch
Journal of Pathology and Bacteriology
Obituary Notice of Deceased Member
Volume 40, Number 3, May 1935

AESTHETICS

Flannery, Maura C.

... although to the non-scientist the aesthetic of biology would mean simply the beauties of nature, to the biologist it means much more. For example, the surface beauty of a leaf is nothing compared to the beauty of its cellular structure and of the process of photosynthesis. Learning about these things just increases appreciation. This is contrary to the idea held by many non-scientists that analysis destroys beauty. This latter view is based on a lack of understanding and knowledge of the processes of science. This is why many of the biologist's beauties are not appreciated by most non-scientists.

Perspectives in Biology and Medicine
Biology is Beautiful (p. 430)
Volume 35, Number 3, Spring 1992

The aesthetic is intrinsic to biology. Biologists are drawn to the field by its aesthetic qualities and continually nurtured by them. This is true in all the sciences, but the aesthetics of biology is a little richer, or at least slightly different.

Perspectives in Biology and Medicine
Biology is Beautiful (p. 433)
Volume 35, Number 3, Spring 1992

AMOEBA

Cudmore, L.L. Larison
Ah, the architecture of this world. Amoebas may not have backbones, brains, automobiles, plastic, television, Valium or any other of the blessings of a technologically advanced civilization; but their architecture is two billion years ahead of its time.

The Center of Life
The Universal Cell (pp. 15–16)

An amoeba never is torn apart through indecision, though, for even if two parts of the amoeba are inclined to go in different directions, a choice is always made. We could interpret this as schizophrenia or just confusion, but it could also be a judicious simultaneous sampling of conditions, in order to make a wise choice of future direction.

The Center of Life
Locomotion (p. 73)

Cuppy, Will
Amoebas not only divide, they also blend. When it's all over there is one amoeba where there were two. Amoebas blend apparently because they enjoy blending for its own sake.

The amoeba often frequents laboratories. You'll find quite a number of amoebas at Yale, Princeton, and Harvard.

How To Get From January to December
March 7 (p. 53)

Huxley, Julian
Amoeba has her picture in the book,
Proud Protozoon!—Yet beware of pride,
All she can do is fatten and divide;
She cannot even read, or sew, or cook...

Essays of a Biologist
Philosophic-Ants (p. 176)

5

Popper, Karl

The difference between the amoeba and Einstein is that, although both make use of the method of trial and error elimination, the amoeba dislikes erring while Einstein is intrigued by it...

Objective Knowledge
Chapter 2, Section 16 (p. 70)

Unknown

An amoeba named Sam and his brother
Were having a drink with each other.
In the midst of their quaffing
They split their sides laughing,
And each of them now is a mother.

Source unknown

When you were a soft amoeba, in ages past and gone,
Ere you were Queen of Sheba, or I King Solomon,
Alone and undivided, we lived a life of sloth,
Whatever you did, I did; one dinner served for both.
Anon came separation, by fission and divorce,
A lonely pseudopodium wandered on my course.

In Arnold Silcock
Verse and Worse
Evolution (pp. 167–8)

AMPHIBIAN

FROG

Carr, Archie
I like the look of frogs, and their outlook, and especially the way they get together in wet places on warm nights and sing about sex.

The Windward Road
The Paradox Frog (p. 90)

Exodus 8:1–4
And Jehovah spake unto Moses, Go in unto Pharaoh, and say unto him, Thus said Jehovah, Let my people go, that they may serve me. And if thou refuse to let them go, behold, I will smite all thy borders with frogs: and the rivers shall swarm with frogs, which shall go up and come into thy house, and into thy bedchambers, and upon thy bed, and into the house of thy servants, and upon thy people, and into thine ovens, and into thy kneading troughs: and the frogs shall come up both upon thee, and upon thy people, and upon all thy servants.

The Bible

Unknown
What a wonderful bird the frog are—
When he stand he sit almost;
When he hop, he fly almost.
He ain't got no sense hardly;
He ain't got no tail hardly either.
When he sit, he sit on what he ain't got almost.

Source unknown

TADPOLE

Kermit the Frog
When I was a tadpole growing up in the swamps, I never imagined that I would one day address such an outstanding group of scholars.

Commencement address
Southampton College
New York, 1996

Ovid
Ev'n slime begets the frog's loquacious race:
Short of their feet at first, in little space
With arms, and legs endu'd, long leaps they take
Rais'd on their hinder parts, and weim the lake,
And waves repel: for Nature gives their kind,
To that intent, a length of legs behind.

In S. Garth (ed.)
Ovid's Metamorphoses, in Fifteen Books
Metamorphoses
Book the Fifteenth (p. 500)

Pallister, William
Three large glass bowls,
In each some half grown tadpoles,
All hatched from the same spawn,

Breathing with gills like fishes
In their small transparent dishes,
Waving their long tails,
Important to themselves as whales,
Some of them to be experimented on;...

Poems of Science
The Nature of Things
Tadpoles (p. 6)

TOAD

Fawcett, Edgar
Blue dusk, that brings the dewy hours,
Brings thee, of graceless form in smooth,
Dark stumbler at the roots of flowers,

Flaccid, inert, uncouth.

In John Burroughs (ed.)
Songs of Nature
A Toad

McArthur, Peter
Probably no creature in all nature has been so villainously libeled as the toad. The greatest of poets speak of "the toad, ugly and venomous," and in fairy lore they are regarded as poisonous. So deeply rooted are these erroneous beliefs that no amount of scientific education seems able to eradicate them. The children are taught in school that the toad is not only harmless, but useful as an insect destroyer, and yet little girls will shriek at a toad just like their mothers.

The Best of Peter McArthur
Toads (p. 177)

Milne, A.A.
(Weasels, Stoats, and Ferrets, together:)
Toad! Toad! Down with Toad!
Down with the popular, successful Toad!

Toad of Toad Hall
Act I, Number 7 (p. 18)

ANALOGY

Chargaff, Erwin
When a science approaches the frontiers of its knowledge, it seeks refuge in allegory or in analogy.

Essays on Nucleic Acids
Chapter 8 (p. 119)

Emerson, Ralph Waldo
...science is nothing but the finding of analogy, identity, in the most remote parts.

The Collected Works of Ralph Waldo Emerson
Volume I
The American Scholar (p. 54)

Hartley, David
Animals are also analogous to Vegetables in many things, and Vegetables to Minerals: So that there seems to be a perpetual Thread of Analogy continued from the most perfect Animal to the most imperfect Mineral, even till we come to elementary Bodies themselves.

Observations on Man
Volume I
Chapter III, Section 1, Proposition 82 (p. 294)

Johnson-Laird, P.N.
A scientific problem can be illuminated by the discovery of a profound analogy, and a mundane problem can be solved in a similar way.

The Computer and the Mind
Chapter 14 (p. 266)

Pepper, Stephen
A man desiring to understand the world looks about for a clue to its comprehension. He pitches upon some area of commonsense fact and

tries to understand other areas in terms of this one. The original area becomes his basic analogy or root metaphor.

World Hypotheses (pp. 91–2)

Strindberg, August

Twice two—is two, and this I will demonstrate by analogy, the highest form of proof. Listen! Once one is one, therefore twice two is two. For that which applies to the one must also apply to the other.

Plays
Dream Play (pp. 561–2)

ANIMAL

Ackerman, Diane
One of the things I like best about animals in the wild is that they're always off on some errand. They have appointments to keep. It's only we humans who wonder what we're here for.

The Moon By Whale Light
Chapter 1 (pp. 41–2)

Agassiz, Louis
Gould, A.A.
Animals are worthy of our regard, not merely when considered as to the variety and elegance of their forms, or their adaptation to the supply

of our wants; but the Animal Kingdom, as a whole, has a still higher signification. It is the exhibition of the divine thought, as carried out in one department of that grand whole which we call Nature; and considered as such, it teaches us most important lessons.

Principles of Zoology
Chapter First (p. 25)

Beston, Henry
We need another and a wiser and perhaps a more mystical concept of animals... We patronize them for their incompleteness, for their tragic fate of having taken form so far below ourselves. And therein we err, and greatly err. For the animal shall not be measured by man. In a world older and more complete than ours they move finished and complete, gifted extensions of the senses we have lost or never attained, living by voices we shall never hear. They are not brethren, they are not underlings, they are other nations, caught with ourselves in the net of life and time, fellow prisoners of the splendor and travail of the Earth.

The Outermost House
Autumn, Ocean, and Birds (p. 25)

Borges, Jorge Luis
...to a certain Chinese encyclopedia entitled *Celestial Emporium of Benevolent Knowledge*... it is written that animals are divided into (a) those that belong to the Emperor, (b) embalmed ones, (c) those that are trained, (d) suckling pigs, (e) mermaids, (f) fabulous ones, (g) stray dogs, (h) those that are included in this classification, (i) those that tremble as if they were mad, (j) innumerable ones, (k) those drawn with a very fine camel's hair brush, (l) others, (m) those that have just broken a flower vase, (n) those that resemble flies from a distance.

Other Inquisitions
The Analytical Language of John Wilkins (p. 103)

Brophy, Brigid
I don't hold animals superior or even equal to humans. The whole case for behaving decently to animals rests on the fact that we are the superior species. We are the species uniquely capable of imagination, rationality, and moral choice—and that is precisely why we are under an obligation to recognize and respect the rights of animals.

Don't Never Forget
The Rights of Animals (p. 21)

Bruchac, Joseph
Let my words
be bright with animals,

images the flash of a gull's wings.
If we pretend
that we are at the center,
that moles and kingfishers,
eels and coyotes
are at the edge of grace,
then we circle, dead moons
almost a cold sun.
This morning I ask only
the blessing of the crayfish,
the beatitude of the birds;
to wear the skin of the bear
in my songs;
to work like a man with my hands.

Prayer
Source unknown

Butler, Samuel
[Wild animals] If one would watch them and know what they are driving at, one must keep perfectly still.

In Geoffrey Keynes and Brian Hill (eds.)
Samuel Butler's Notebooks
Wild Animals and One's Relations (p. 112)

Canetti, Elias
Whenever you observe an animal closely, you feel as if a human being sitting inside were making fun of you.

The Human Province
1942 (p. 7)

Ehrlich, Gretel
Animals give us their constant, unjaded faces and we burden them with our bodies and civilized ordeals.

The Solace of Open Spaces
Friends, Foes, and Working Animals (p. 62)

Eiseley, Loren
Animals are molded by natural forces they do not comprehend. To their minds there is no past and no future. There is only the everlasting present of a single generation—its trails in the forest, its hidden pathways in the air and in the sea.

The Star Thrower
The Long Loneliness (p. 37)

Eliot, George
Animals are such agreeable friends—they ask no questions, they pass no criticism.

Scenes of Clerical Life
Mr Gilfil's Love Story
Chapter VII (p. 129)

Gardner, John
Always be kind to animals,
Morning, noon, and night;
For animals have feelings too,
And furthermore, they bite.

A Child's Bestiary
Introduction

Krutch, Joseph Wood
... the most important reason why there are so many gaps in the available life histories of even the commoner animals is less the perversity of professors than the fact that there are an awful lot of these common creatures and that actually to follow their lives from day to day is a very difficult time-consuming task.

The Desert Year
The Contemplative Toad (p. 109)

We have never entered into an animal's mind and we cannot know what it is like, or even if it exists. The risk of attributing too much is no greater than the risk of attributing too little.

The Great Chain of Life
Prologue (p. x)

Oken, Lorenz
Animal is blossom without a stem.

In H.R. Hays
Birds, Beasts, and Men
Chapter 17 (p. 212)

Poe, Edgar Allan
There is something in the unselfish and self-sacrificing love of a brute which goes directly to the heart of him who has had frequent occasion to test the paltry friendship and gossamer fidelity of mere *Man*.

Little Masterpieces
The Black Cat (p. 128)

Pratchett, Terry
[For animals] the whole panoply of the universe has been neatly expressed to them as things to (a) mate with, (b) eat, (c) run away from, and (d) rocks.

Equal Rites (p. 78)

Purcell, Rosamond
Gould, Stephen Jay
Animals in nature, contrary to the suspicions of cynics or the hopes of idealists, are neither intrinsically vicious nor altruistic. Competition and cooperation are both nature's ways.

Illuminations: A Bestiary (p. 101)

Sanborn, Kate
...if Darwin's theory should be true, it will not degrade man; it will simply raise the whole animal world into dignity, leaving man as far in advance as he is at present.

Atlantic Monthly
Studies of Animal Nature (p. 135)
February, 1877

AARDVARK

Unknown
...it's aardvark, but it pays well.

In John S. Crosbie
Crosbie's Dictionary of Puns (p. 5)

APE

Young, Roland
The sacred ape, now, children, see.
He's searching for the modest flea.
If he should turn around we'd find
He has no hair on his behind.

Not for Children
The Ape

ARMADILLO

Nash, Ogden
The armadillo lives inside
A corrugated plated hide.
Below the border this useful creature
Of tidy kitchens is a feature,
For housewives use an armadillo
To scour their pots, instead of Brillo.

Everyone But Thee and Me
The Armadillo

ASS

Goldsmith, Oliver
JOHN TROTT was desired by two witty Peers
To tell them the reason why asses had ears?
"An't please you," quoth John, "I'm not given to letters,
Nor dare I pretend to know more than my betters;
Howe'er, from this time I shall ne'er see your *graces*,
As I hope to be saved! without thinking on *asses*."

The Complete Poetical Works of Oliver Goldsmith
The Clown's Reply

BAT

Berryman, John
Bats have no banks and they do not drink and cannot be arrested and pay
no tax and, in general; bats have it made.

77 Dream Songs
Number 63

Dawkins, Richard
A bat is a machine, whose internal electronics are so wired up that its wing muscles cause it to home in on insects, as an unconscious guided missile homes in on an aeroplane.

The Blind Watchmaker
Chapter 2 (p. 37)

Montgomery, James
What shall I call thee—bird, or beast, or neither?
—Just what you will; I'm rather both than neither;
Much like the season when I whirl my flight,
The dusk of evening,—neither day nor night.

Poetical Works of James Montgomery
Volume II
The Bat

Nash, Ogden
Myself, I rather like the bat,
It's not a mouse, it's not a rat.
It has no feathers, yet has wings,
It's quite inaudible when it sings.
It zigzags through the evening air
And never lands on ladies' hair,
A fact of which men spend their lives
Attempting to convince their wives.

Verses from 1929 On
The Bat

Tabb, John Banister
To his cousin the Bat
Squeaked the envious Rat,
"How fine to be able to fly!"
Tittered she, "Leather wings
Are convenient things;
But nothing *to sit on* have I."

The Poetry of Father Tabb
Humerous Verse
An Inconvenience

BEAR

Lear, Edward
There was an old person of Ware,
Who rode on the back of a bear;
When they ask'd, "Does it trot?" he said, "Certainly not!
He's a Moppsikon Floppsikon bear!"

Of Pelicans and Pussycats

Pope, Alexander
The fur that warms a monarch, warm'd a bear.

Alexander Pope's Collected Poems
Essay on Man
Epistle III, l. 44

BEAVER

Outwater, Alice
The beaver is utterly familiar. Forty inches long and over a foot upright, a beaver seems like a little person with a fondness for engineering.

Water: A Natural History
Chapter 2 (p. 19)

BUFFALO

Unknown
The buffalo is the death
that makes a child climb a thorn tree.
...
He is the butterfly of the savannah:
He flies along without touching the grass.
When you hear thunder without rain—
it is the buffalo approaching.

In Ulli Beier
Yoruba Poetry
Buffalo

CAT

Krutch, Joseph Wood
... cats seem to go on the principle that it never does any harm to ask for what you want.

The Twelve Seasons
February (p. 160)

Whitehead, Alfred North
If a dog jumps in your lap, it is because he is fond of you; but if a cat does the same thing, it is because your lap is warmer.

In Lucien Price
The Dialogues of Alfred North Whitehead
Chapter XXV
December 10, 1941 (p. 187)

CENTIPEDE

Blanshard, Brand
The centipede was happy quite
Until the toad for fun
Said: 'Pray which leg comes after which?'
This wrought his mind to such a pitch,
He lay distracted in the ditch
Considering how to run.

The Nature of Thought
Volume 1
Chapter VI, fn 1 (p. 232)

CHIMPANZEE

Herford, Oliver
Chil-dren, be-hold the Chim-pan-zee:
He sits on the an-ces-tral tree
From which we sprang in ag-es gone.
I'm glad we sprang: had we held on,
We might, for aught that I can say,
Be hor-rid Chim-pan-zees to-day.

A Child's Primer of Natural History
The Chimpanzee

CORAL

Crabbe, George
Involved in sea-wrack, here you find a race,
Which science, doubting, knows not where to place;
On shell or stone is dropp'd the embryo-seed,
And quickly vegetates a vital breed.

Poems
Volume I
The Borough
Letter IX, l. 90–4

Day, Richard Edwin
Out of the gardens of the deep,
Out of the orchards of the sea—
Farther than ever storm-keels sweep—
Blossomed the coral tree.

Poems
The Coral Tree

Gerhard, John
Although Corrall be a matter or substance, even as hard as stones; yet I
think it not amisse to place and insert it here next unto the mosses, and
the rather for that the kindes therof do shew themselves, as well in the
maner of their growing, as in their place and some, like unto the Mosses.

The Herball or Generall Historie of Plantes
Book 3
Chapter 166 (p. 1576)

COW

Young, Roland
The cow's a gentle, patient soul,
With milk she fills the flowing bowl.
She's kind to babies, mean to flies,
She has the most coquettish eyes.

Not for Children
The Cow

COYOTE

James, William
I saw a moving sight the other morning before breakfast... The young man
of the house had shot a little wolf called a coyote in the early morning.

The heroic little animal lay on the ground, with his big furry ears, and his clean white teeth, and his jolly cheerful little body, but his brave little life was gone. It made me think how brave all these living things are. Here little coyote was, without any clothes or house or books or anything, with nothing but his own naked self to pay his way with, and risking his life so cheerfully—and losing it—just to see if he could pick up a meal near the hotel. He was doing his coyote-business like a hero...

In Henry James (ed.)
The Letters of William James
Volume II
Letter to his Son Alexander
August 28, 1898 (pp. 81–2)

DOG

Butler, Samuel
The greatest pleasure of a dog is that you may make a fool of yourself with him and not only will he not scold you, but he will make a fool of himself.

In Geoffrey Keynes and Brian Hill (eds)
Samuel Butler's Notebooks
Dog (p. 314)

Nash, Ogden
The truth I do not stretch or shove
When I state the dog is full of love.

I've also proved by actual test,
A wet dog is the lovingest.

Everyone But Thee and Me
The Dog (p. 71)

DUCK-BILLED PLATYPUS

Flanders, Michael
Minale, Marcello
We call him "Duck-billed Platypus"
And mock him for his name:
He does not seem to mind it.
He feels no sense of shame
Because he does not know himself
By such a title,
He's
A "Golden, Shining Love-Bird"
In Duck-billed Platypese.

Creatures Great and Small...
The Duck-billed Platypus

ELEPHANT

Cuppy, Will
In the Pleistocene Era, there were more than twenty kinds of elephants.
Now there are only two. That's plenty.

How to Get From January to December
April 24 (p. 85)

Donne, John
Nature's great masterpiece, an Elephant.
The only harmless great thing; the giant of beasts.

Complete English Poems
The Progress of the Soul
Stanza 39

Shakespeare, William
The elephant hath joints, but none for courtesy. His legs are legs for necessity, not for flexure.

Troilus and Cressida
Act II, scene III, l. 113

Swift, Jonathan
So Geographers in *Afric*-Maps
With Savage Pictures fill their gaps;
And o'er unhabitable Downs
Place Elephants for want of Towns.

On Poetry
A Rhapsody, l. 177–80

EUGLENA VIRIDIS

Pallister, William
A plant when there is sunshine; an animal at night.
The living proof of theories, biologists' delight,
Created by environment and matching it so well,
You are both plant and animal. Which one the time can tell.

Poems of Science
Euglena Viridis

GIRAFFE

Young, Roland
Now, children, you must never laugh
At the stately tall giraffe.
She's sensitive, as you can tell;
But, my dears, she kicks like hell!

Not for Children
The Giraffe

GORILLA

Bradley, Mary Hastings
The gorilla is a strict vegetarian like the elephant and buffalo—three of the four most dangerous animals in Africa. It behooves one to walk softly with vegetarians.

On the Gorilla Trail
Chapter IX (p. 131)

GUANACO

Simpson, George Gaylord
The guanaco is a camel but
He hasn't got a hump.

He's about three-quarters mountain goat
And seven-eighths a chump.

Concession to the Improbable
Chapter 8 (p. 72)

HIPPOPOTAMUS

Belloc, Hilaire
I shoot the Hippopotamus
With bullets made of platinum,
Because if I use leaden ones
His hide is sure to flatten 'em.

Complete Verse
The Hippopotamus

Macaulay, Thomas Babington
I have seen the hippopotamus, both asleep and awake; and I can assure you that, awake or asleep, he is the ugliest of the works of God.

Letter to Macvey Nappier
March 9, 1850

HORSE

Shakespeare, William
A horse! A horse! My kingdom for a horse!

The Tragedy of King Richard the Third
Act V, scene IV, l. 7

Twain, Mark
I have known the horse in war and in peace, and there is no place where a horse is comfortable. The horse has too many caprices, and he is too much given to initiative. He invents too many ideas. No, I don't want anything to do with a horse.

The Complete Works of Mark Twain
Mark Twain's Speeches
Welcome Home (p. 201)

JACKAL

Byron, George
The jackal's troop, in gather'd cry,
Bay'd from afar complainingly,

With a mix'd and mournful sound,
Like crying babe, and beaten hound.

The Complete Poetical Works
Volume III
Siege of Corinth
Stanza 33, l. 1024–7

JELLY-FISH

Kendall, May
Her beauty, passive in despair,
Through sand and seaweed shone,
The fairest jelly-fish I e'er
Had set mine eyes upon.

It would have made a stone abuse
The callousness of fate,
This creature of prismatic hues,
Stranded and desolate!

Dreams to Sell
The Philanthropist and the Jelly-Fish

Allen, Grant
A jellyfish swam in a tropical sea,
And he said, "This world it consists of me:
There's nothing above and nothing below
That a jellyfish ever can possibly know
(Since we've got no sight, or hearing, or smell),
Beyond what our single sense can tell.

In E. Haldeman-Julius
Poems of Evolution
The First Idealist

LEOPARD

Wells, Carolyn
If strolling forth, a beast you view,
Whose hide with spots is peppered,
As soon as he has lept on you,
You'll know it is the leopard.
'Twill do no good to roar with pain,
He'll only lep and lep again.

Baubles
How to Tell the Wild Animals

LION

Gay, John
The Lion is (beyond dispute)
Allow'd the most majestic brute;
His valour and his gen'rous mind
Prove him superior of his kind.

The Poetical Works of John Gay
Volume II
The Fables
Volume the Second
Fable IX
The Jackal, Leopard, and Other Beasts

Pringle, Thomas
Wouldst thou view the Lion's den?
Search afar from haunts of men,—
Where the reed-encircled rill,
Oozes from the rocky hill,
By its verdure far descried
'Mid the desert brown and wide.

Afar in the Desert
The Lion and the Giraffe
Stanza 1

LLAMA

Belloc, Hilaire
The Llama is a woolly sort of fleecy hairy goat,
With an indolent expression and undulating throat
Like an unsuccessful literary man.

Complete Verse
The Llama

Nash, Ogden
The one-l lama,
He's a priest.
The two-l llama,
He's a beast.
And I will bet
A silk pajama
There isn't any three-l lllama.

Verses from 1929 On
The Lama

MANATEE

Nash, Ogden
The manatee is harmless
And conspicuously charmless.
Luckily the manatee
Is quite devoid of vanity.

Verses from 1929 On
The Manatee

MILLIPEDE

Garstang, Walter
The hatching of a Millipede brings curious things to light:
The embryo within its shell is curled up snug and tight
Enclosed inside an inner skin with a thorn upon its neck,
Whose task it is to pierce the shell, as chicks their prisons peck.
What is this extra covering that thus comes into view?
An heirloom from antiquity here blended with the new?
Another "Nauplius-coat" around another embryo,
The same that Peracarids on their cradled babes bestow?

Larval Forms
The Millipede's Egg-Tooth
Stanza 1 (p. 51)

MOUSE

Cuppy, Will
I have nothing against mice, in moderation…My own mice just eat whatever I have in the place, including soap. Not an ideal diet, but they'll have to make it do or move elsewhere.

How to Get From January to December
January 20

Unknown
The goal of science is to build better mousetraps. The goal of nature is to build better mice.

Source unknown

OTTER

Colum, Padraic
I'll be an otter, and I'll let you swim
A mate beside me; we will venture down
A deep, full river when the sky above
Is shut of the sun; spoilers are we;
Thick-coated; no dog's tooth can bite at our veins—
With ears and eyes of poachers; deep-earthed ones
Turned hunters; let him slip past,
The little vole, my teeth are on an edge
For the King-fish of the River!

Poems
Otters

PANDA

Schaller, George B.
Jinchu, Hu
Wenshi, Pan
Jing, Zhu
There are two giant pandas, the one that exists in our mind and the one that lives in its wilderness home. Soft, furry, and strangely patterned in black and white, with a large, round head and a clumsy, cuddly body, a panda seems like something to play with and hug. No other animal has so entranced the public...The real panda, however, the panda as it lives in the wild, has remained essentially a mystery.

The Giant Pandas of Wolong
Introduction (p. xiii)

PANTHER

Bierce, Ambrose
Lifting her eyes she saw two bright objects starring the darkness with a reddish-green glow. She took them to be two coals on the hearth, but with her returning sense of direction came the disquieting consciousness that they were not in that quarter of the room, moreover were too high, being nearly at the level of the eyes—of her own eyes. For these were the eyes of a panther.

The Eyes of the Panther
The Eyes of the Panther (pp. 18–19)

PECCARY

Wilson, Edward O.
A tame peccary watched me with beady concentration from beneath the shadowed eaves of a house. With my own, taxonomist's eye I registered the defining traits of the collared species, *Dicotyles tajacu*: head too large for the piglike body, fur coarse and brindled, neck circled by a pale thin stripe, snout tapered, ears erect, tail reduced to a nub. Poised on still little dancer's legs, the young male seemed perpetually fierce and ready to charge yet frozen in place, like the metal boar on an ancient Gallic standard.

Biophilia
Bernhardsdorp (p. 4)

PIG

Perrin, Noel
Pigs get bad press. Pigs are regarded as selfish and greedy—as living garbage pails. Pigs are the villains in George Orwell's *Animal Farm*. Pigs have little mean eyes. There is truth in this account—not that it's entirely the fault of the pigs. For perhaps five thousand generations pigs have been deliberately bred to be gluttonous... Do the same thing with human beings for five thousand generations, and it would be interesting to see what kind of people resulted.

Second Person Rural
Pig Tales (p. 143)

POLAR BEAR

Belloc, Hilaire
The Polar Bear is unaware
Of cold that cuts me through:
For why? He has a coat of hair,
I wish I had one too!

Complete Verse
The Polar Bear

PORPOISE

Twain, Mark
The porpoise is the kitten of the sea: he never has a serious thought, he cares for nothing but fun and play.

Following the Equator
Volume I
Chapter IX (p. 110)

PRAIRIE-DOG

Austin, Mary
Old Peter Prairie-Dog
Builds him a house
In Dog-Dog Town,
With a door that goes down
And down and down,
And a hall that goes under
And under and under,
Where you can't see the lightning,
You can't hear the thunder,
For they don't *like* thunder
In Dog-Dog Town.

The Children Sing in the Far West
Dog-Dog Town

RHINOCEROS

Belloc, Hilaire
Rhinoceros, your hide looks all undone,
You do not take my fancy in the least:
You have a horn where other brutes have none:
Rhinoceros, you are an ugly beast.

Complete Verse
The Rhinoceros

SEA SQUIRT

Dennett, Daniel C.
The juvenile sea squirt wanders through the sea searching for a suitable rock or hunk of coral to cling to and make its home for life. For this task, it

has a rudimentary nervous system. When it finds its spot and takes root, it doesn't need its brain anymore so it eats it! (It's rather like getting tenure.)

Consciousness Explained
Chapter 7 (p. 177)

SHREW

Huxley, Julian
Timid atom, furry shrew,
Is it a sin to prison you?
Through the runways in the grass
You and yours in hundreds pass,
An unimagined world of shrews,
A world whose hurrying twilight news
Never stirs but now and then
The striding world of booted men.

The Captive Shrew
The Captive Shrew

Schaefer, Jack
Shrews are not mutual murderers. We'll just square off and touch whiskers, assessing each other. Then we'll try to out-squeak each other.

Audubon
Interview with a Shrew (p. 2)
Volume 77, Number 6, November 1975

SKUNK

Young, Roland
In this mechanic age the skunk
Inspires no terror—he's the bunk;
For people in cars,
Returning from bars,
Quite frequently flatten the skunk.

Not for Children
The Skunk

SLUG

Deyrup, Olsen, Ingrith
Most people think, "Slugs—yuk!" But I think that whenever you start to study an organism, you become overwhelmed by the beauty and

complexity of it. I am always *amazed* and *touched* by the way these animals solve the tremendous problems they have, which are always really basically the same as ours. I have come to have *very* strong respect and admiration for them, and I've also found it's a *wonderful* area to involve nonscientists in. The minute you begin to show them that slugs are very complicated, interesting animals with their own needs and demands, people begin to look at them with very different eyes. I'm very moved by the slug's ingenuity and tremendous drive to continue living. I think in the end this is what makes me go on, no matter *how* frustrating the experiments happen to be at that time.

In Linda Jean Shepherd
Lifting the Veil
Chapter 3 (pp. 69–70)

SPONGE

Gerhard, John

There is found growing upon the rocks near the sea, a certain matter wrought together of foame or froth of the sea which we call sponges.

The Herball or Generall Historie of Plantes
Book 3
Chapter 166 (p. 1578)

SQUIRREL

Prelutsky, Jack

Squirrels, often found in parks,
have tails resembling question marks,
it's just coincidental, though...
there's little squirrels care to know.

Something Big Has Been There
Squirrels

TIGER

Lawrence, D.H.

I consider the tiger as a *being*, a created being. If you kill all tigers still the tiger-soul continues... But the point is I don't *want* the tiger superseded. Oh, may each she-tigress have seventy seven whelps, and may they all grow in strength and shine in stripes like day and night, and may each one eat at least seventy miserable featherless human birds, and lick red chops of gusto after it.

In James T. Boulton
Selected Letters of D.H. Lawrence
Chapter III
May 1921
Letter to Earl and Achsah Brewster
15 May 1921 (pp. 204, 205)

Wells, Carolyn
Or if some time when roaming round,
A noble wild beast greets you,
With black stripes on a yellow ground,
Just notice if he eats you.
This simple rule may help you learn
The Bengal tiger to discern.

Baubles
How to Tell Wild Animals

WALRUS

Carroll, Lewis
'The time has come,' the Walrus said,
'To talk of many things:
Of shoes—and ships—and sealing-wax—
Of cabbages—and kings—
And why the sea is boiling hot—
And whether pigs have wings.'

The Complete Works of Lewis Carroll
Through the Looking-Glass
Chapter IV (p. 186)

WHALE

Chief Engineer Scott
Admiral, there be whales here!

Star Trek IV
The Voyage Home

WHELK

Wood, Robert William
...if you listen to the shell,

In which the Whelk is said to dwell,
And hear a roar, beyond a doubt
It indicates the Whelk is out.

How to Tell the Birds from the Flowers and Other Woodcuts
The Elk. The Whelk. (p. 43)

YAK

Belloc, Hilaire
As a friend to the children commend me the Yak.
You will find it exactly the thing:
It will carry and fetch, you can ride on its back,
Or lead it about with a string.

Complete Verse
The Yak

Smith, William Jay
The long-haired Yak has long black hair,
He lets it grow—he doesn't care.
He lets it grow and grow and grow,
He lets it trail along the stair.
Does he ever go to the barbershop? NO!
How wild and woolly and devil-may-care
A long-haired Yak with long black hair
Would look when perched in a barber chair!

Mr Smith and Other Nonsense
Yak

ARACHNIDS

MITES

Duck, Stephen
Dear Madam, did you never gaze
Thro' optic glass on rotten cheese?
There, Madam, did you ne'er perceive
A crowd of dwarfish creatures live?
The little things, elate with Pride,
Strut to and fro, from side to side:
In tiny pomp and partly vein,
Lords of their pleasing orb they reign;
And fill'd with harden'd Curds and Cream,
Think the Whole Dairy made for them.

In T.E. Hughes
Mites, or the Acari (p. vii)

Frost, Robert
A speck that would have been beneath my sight
On any but a paper sheet so white
Set off across what I had written there.
And I had idly poised my pen in air
To stop it with a period of ink
When something strange about it made me think.
This was no dust speck by my breathing blown,
But unmistakably a living mite
With inclinations it could call its own.

The Poetry of Robert Frost
A Considerable Speck

Hooke, Robert
The least of *Reptiles* I have hitherto met with, is a Mite.

Micrographia
Observation LV (p. 213)

Unknown
The cheese-mites asked how the cheese got there,
And warmly debated the matter;
The orthodox said it came from the air,
And the heretics said from the platter.

In Arnold Silcock
Verse and Worse
Four More Brief Beliefs (p. 60)

SCORPION

Belloc, Hilaire
The Scorpion is as black as soot,
He dearly loves to bite;
He is a most unpleasant brute
To find in bed, at night.

Complete Verse
The Scorpion

SPIDERS

Dickinson, Emily
The spider as an artist
Has never been employed
Though his surpassing merit
Is freely certified

By every broom and Bridget
Throughout a Christian land.
Neglected son of genius,
I take thee by the hand.

Poems By Emily Dickinson
XCV

Flanders, Michael
Swann, Donald
I have fought a grizzly bear,
Tracked a cobra to its lair,
Killed a crocodile who dared to cross my path;
But the thing I really dread
When I've just got out of bed

Is to find that there's a spider in the bath.

In Paul Hillyard
The Books of the Spider
Driven To It—By the Spider in the Bath! (p. 37)

Florian, Douglas
O Daddy
Daddy O
How'd you get
Those legs to grow
So very long
And lean in size?
From spiderobic
Exercise?

Insectlopedia
The Daddy Longlegs

Pallister, William
Of the SPIDERS and SCORPIONS, five thousand kinds:
These are scattered abroad, on the sea and the shore,
Quite unpleasant to think of, but still it reminds
To be glad there are not many thousand kinds more.
These are eight-legged beauties, with schemes of their own,
And the safest precaution is: Leave them alone!

Poems of Science
Beginnings
Animal Life (p. 140)

Smith, Bertha Wilcox
Throughout the night he spun a thread
With which he wove medallioned lace
That stretched between two milkweed pods
Beside a dusty, traveled place;
The pattern was a scalloped round—
Each radius exactly drawn
With trellised filaments between,
And over all bright diamonds shone;
In meshed and tenuous design
It was a fragile, wayside sonnet—
The maker, heedless of acclaim,
Had left no signature upon it.

Nature Magazine
Anonymous (p. 234)
Volume 50, Number 5, May 1957

Taylor Family
'O look at that great ugly Spider,' said Ann,
And screaming, she knocked it away with her fan;
"T is a great ugly creature, as ever can be,
I wish that it would not come crawling on me.'

Original Poems for Infant Minds
The Spider
Stanza I

White, Terence Hanbury
A spider is an air worm, as it is provided with nourishment from the air,
which a long thread catches down to its small body.

The Book of Beasts (p. 191)

TICKS

Florian, Douglas
Not gigan-tic.
Not roman-tic.
Not artis-tic.
Not majes-tic.
Not magne-tic.
Nor aesthe-tic.
Ticks are strictly parasi-tic.

Insectlopedia
The Ticks

BACTERIA

Cohn, Ferdinand

If one could inspect a man under a similar lens-system he would appear as big as Mont Blanc or even as Mt Chimborazo. But even under these colossal magnifications the smallest bacteria look no larger than the periods and commas of good print; little or nothing can be distinguished of their inner parts, and of them their very existence would have remained unsuspected if it had not been for their countless numbers.

In Kenneth Thimann
The Life of Bacteria
Chapter II (p. 33)

Helmuth, W.T.

Oh, powerful bacillus,
With wonder how do you fill us,
   Every day!
While medical detectives,
With powerful objectives,
   Watch you play.

Ode to the Bacillus
Source unknown

BACTERIOLOGIST

Esar, Evan
A bacteriologist is a man whose conversation always start with the germ of an idea.

20,000 Quips & Quotes

BEAUTY

Awiakta, Marilou
Beauty is no threat to the wary
  who treat the mountain in its way,
    the copperhead in its way,
      and the deer in its way,
knowing that nature is the human heart
made tangible.

Selu: Seeking the Corn-Mother's Wisdom
Trail Warning (p. 39)

da Vinci, Leonardo
Even though the genius of man might make various inventions, attaining the same end by various means, it will not invent anything more beautiful, or more economical, or more direct than nature, for in nature's inventions nothing is wanting and nothing is superfluous.

Cold Spring Harbor Symposia on Quantitative Biology
Quoted in Theodosius Dobzhansky
Evolution of Genes and Genes in Evolution (p. 15)
Volume XXIV, 1959

Leibniz, Gottfried Wilhelm
The beauty of nature is so great and its contemplation so sweet... whoever tastes it, can't help but view all other amusements as inferior.

In Ernst Peter Fischer
Beauty and the Beast
Chapter 2 (p. 47)

Weil, Simone
The true subject of science is the beauty of the world.

In Ernst Peter Fischer
Beauty and the Beast
Chapter 5 (p. 91)

BIOLOGICAL

Arber, Agnes
Since the first step in biological research involves the decision as the question on which to concentrate, the researcher is at once put upon his mettle, for the full recognition and appreciation of a problem may task him even more severely than its solution.

The Mind and the Eye
Chapter I (p. 6)

Bernard, Claude
If we mean to build up the biological sciences, and to study fruitfully the complex phenomena which occur in living beings, whether in the physiological or the pathological state, we must first of all lay down principles of experimentation, and then apply them to physiology, pathology and therapeutics.

An Introduction to the Study of Experimental Medicine
Introduction (p. 2)

Bird, J.M.
...we shall have to have a philosophy of biological life which gives the human animal something to survive with, a universe which gives us a place to survive into, and a covering of cosmic philosophy which recognizes all this as an aspect of reality. If the necessity arises it will be met and in that event we shall be able to say with obvious truth that science and religion have come together.

In E.H. Cotton
Has Science Discovered God?
Chapter XVI (p. 293)

Brower, David
A fallen tree supports a biological community that may be essential to the existence of the forest itself.

In Jonathan White
Talking on the Water
The Archdruid Himself (p. 41)

Chargaff, Erwin
An observer of our biological sciences today sees dark figures moving over a bridge of glass. We are faced with an ever expanding universe of light and darkness. The greater the circle of understanding becomes, the greater is the circumference of surrounding ignorance.

Essays on Nucleic Acids
Chapter 8 (p. 109)

Compton, Karl Taylor
More recently in the development of a program of biological engineering, based upon physical, chemical, and biological operations, a similar attempt has been made to synthesize an appropriate training for the handling of a great variety of biological situations, whether they be in the food industry or in the hospital or medical or biological research fields. I suspect that there may be other directions in which an analogous approach may be made to simplify the educational program and at the same time increase the power acquired by the student.

A Scientist Speaks (p. 53)

Dayton, P.K.
Mordida, B.J.
Bacon, F.
Geological history and oceanographic processes are the warp and woof of the biological understanding of any marine habitat.

American Zoologist
Polar Marine Communities (p. 90)
Volume 34, 1994

Dunn, R.A.
Davidson, R.A.
Biologic categorization is one of the most conspicuous aspects of successful behavior, not only of man, but of all animals, in meeting the requirements for survival in a complex environment.

Pattern Recognition
Pattern Recognition in Biological Classification (p. 75)
Volume 1, 1968

Durant, Will
Durant, Ariel
So the first biological lesson of history is that life is competition. Competition is not only the life of trade, it is the trade of life—peaceful when food abounds, violent when the mouths outrun the food. Animals eat one another without qualm; civilized men consume one another by due process of law.

The Lessons of History
Chapter III (p. 19)

Handler, Philip
Biology has become a mature science as it has become precise and quantifiable. The biologist is no less dependent upon his apparatus than the physicist.

Biology and the Future of Man
Chapter 1 (p. 6)

Loewy, A.G.
Siekevitz, P.
A dramatic demonstration of the importance of biological structure was provided by the experiments of Skoultchi and Morowitz, who cooled the eggs of the brine shrimp Artemia to temperatures below -271 centigrade and showed that upon rewarming their hatch rate was the same as that of control eggs held at room temperature. Since at that temperature we have structure but presumably no process, it is reasonable to conclude that structure is not only a necessary condition, but even a sufficient condition for initiating biological function. It would thus appear that living processes could be generated by putting together the proper structures, the synthesis of life becoming "merely" a very complicated exercise in organic chemistry.

Cell Structure and Function
Chapter 4 (p. 33)

Pittendrigh, Colin S.
The study of adaptation is not an optional preoccupation with fascinating fragments of natural history, it is the core of biological study.

In A. Roe and G.G. Simpson (eds)
Behavior and Evolution
Adaptation, Natural Selection, and Behavior (p. 395)

Snyder, Gary

We're so impressed by our civilization and what it's done, with our machines, that we have a difficult time recognizing that the biological world is infinitely more complex.

The Real Work
Tracking Down the Natural Man (p. 87)

Trivers, Robert

I want to change the way people think about their everyday lives. How you think is going to affect who you marry, what kind of relationship you establish, whether and in what manner you reproduce. That's day-to-day thinking, right? But they don't even teach courses on that stuff...Life *is* intrinsically biological. It's absurd not to use our best biological concept.

In Roger Bingham
A Passion to Know: 20 Profiles in Science
Robert Trivers: Biologist of Behavior (p. 75)

Unknown

But people have a right to ask why biologists should tinker with these things at all. Well, we all want to have our cancers cured, our diseased organs replaced, our congenital deformities prevented, our food supplies assured, our epidemics controlled, our nerves soothed; and if we want these things we shall simply have to take the risk of scientists doing the experiments they need to do to achieve these ends. Society cannot, and should not try to, decide what a man may discover, but society can and does decide how any discovery is to be applied...In all highly developed communities elaborate machinery exists to determine and control expenditure on scientific research; and even the smallest projects are carefully scrutinized by competent and socially responsible bodies before they are accepted. There is not the slightest chance of any biological discovery being applied to society as a whole unless society wants it to be applied.

Review of G.R. Taylor
The Biological Time-Bomb
in *The Times Literary Supplement*
25 April 1968

Wesenberg-Lund, C.

From a purely scientific point of view, I have always regarded the question, who first made a biological observation, as a matter of sublime indifference. It must never be forgotten that even with regard to biological observations, which can only rarely be committed to paper with the same convincing exactness as an anatomical structure, the exact apprehension of a given fact can only be acquired through repeated observation. It is further of the greatest significance that the biological observations are tested by different scientists and in different latitudes; only in that way can our suppositions and hypotheses be registered among real scientific facts. It must further be remembered that the study of Nature must always begin with the slightest possible literary ballast. He who has first crammed his head with all that has been written upon a subject will, at the moment of observation, when standing face to face with Nature, soon understand that his whole learning is only felt as a burden and restricts his power of observation. I for my own part have always been of the opinion that it is exactly the smallest equipment of human knowledge which gives the greatest peace in my studies, creates the scientific sovereignty over observations and thoughts and—as far as possible—moves the milestones of time nearer to the borders of eternity.

In Marston Bates
The Natural History of Mosquitoes
Chapter XIX (p. 286)

Wheeler, William Morton

And so far as the actual, fundamental, biological structure of our society is concerned and notwithstanding its stupendous growth in size and all the tinkering to which it has been subjected, we are still in much the same infantile stage. But if the ants are not despondent because they have failed to produce a new social invention or convention in 65 million years, why should we be discouraged because some of our institutions and castes have not been able to evolve a new idea in the past fifty centuries?

Social Life Among the Insects
Lecture I (pp. 8–9)

Woodger, Joseph Henry

If we make a general survey of biological science we find that it suffers from cleavages of a kind and to a degree which is unknown in such a well unified science as, for example, chemistry. Long ago it has undergone that inevitable process of subdivision into special branches which we find in other sciences, but in biology this has been accompanied by a characteristic divergence of method and outlook between the exponents of the several branches which has tended to exaggerate their differences and has even led to certain traditional feuds between them. This process of fragmentation continues, and with it increases the time and labour requisite for obtaining a proper acquaintance with any particular branch.

Biological Principles: A Critical Study
General Introduction (p. 11)

Young, Michael
Every bodily process is pulsing to its own beat within the overall beat of
the solar system.

The Metronomic Society
Chapter 2 (p. 20)

BIOLOGIST

Connolly, Cyril

The answer seems to rest with three categories of thinkers; the physicists, who incline to believe in God but are now all busy making explosives; the biologists and chemists who can produce almost everything except life and who, if they could create life, would prove that it might once have arisen accidentally; and the psychologists and physiologists, who are struggling to discover the relation of mind to brain, the nature of consciousness.

The Unquiet Grave
Part III (pp. 106–7)

Cudmore, L.L. Larison

All cell biologists are condemned to suffer an incurable secret sorrow: the size of the objects of their passion.

The Center of Life
The Universal Cell (p. 5)

Flannery, Maura C.

The patterns and rhythms of nature, science as a search for order, form as a central problem in biology, are themes that are rarely emphasized in research reports and in texts, they are nevertheless powerful concepts that direct and inform biologists' work.

Perspectives in Biology and Medicine
Biology is Beautiful (p. 427)
Volume 35, Number 3, Spring 1992

Hull, D.L.

Evolutionary biologists are currently confronted by a...dilemma: If they insist on formulating evolutionary theory in terms of commonsense entities, the resulting laws are likely to remain extremely variable and complicated; if they want simple laws, equally applicable to all entities of a particular sort, they must abandon their traditional ontology. This reconceptualization of the evolutionary processes is certainly counter-intuitive; its only justification is the increased scope, consistency, and power of the theory that results.

Annual Review of Ecology and Systematics
Individuality and Selection (pp. 316–17)
Volume 11, 1980

Huxley, Thomas H.

I do not question for a moment, that while the Mathematician is busied with deductions *from* general propositions, the Biologist is more especially occupied with observations, comparisons, and those processes which lead *to* general propositions.

Lay Sermons, Addresses, and Views
On the Educational Value of the Natural History Sciences (p. 87)

Kellog, Vernon

...the biologist seems unable to escape from the use of a terminology that is to be found in the larger dictionaries—and these dictionaries are at home, while the public is in the lecture-hall.

The Atlantic Monthly
The Biologist Speaks of Death (p. 778)
June 1921

Leob, Jacques

...the investigations of the biologist differ from those of the chemist and physicist in that the biologist deals with the analysis of the mechanism of a special class of machines. Living organisms are chemical machines, made of essentially colloidal material which possess the peculiarity of developing, preserving and reproducing themselves automatically. The machines which have thus far been reproducing themselves, though no one can say with certainty that such machines might not one day be constructed artificially.

Science
The Recent Development of Biology (p. 777)
Volume 20, 1904

Medawar, Peter

Biologists work very close to the frontier between bewilderment and understanding. Biology is complex, messy and richly various, like real life; it travels faster nowadays than physics or chemistry (which is just as well, because it has so much farther to go), and it travels nearer to the ground.

Pluto's Republic
Induction and Intuition in Scientific Thought (p. 73)

Salthe, Stanley N.
...we are, as evolutionary biologists, indirectly working on *nothing less than an important part of our culture's very own creation myth.* Is the combination of the pointlessness of chance with the tyranny of necessity, competitive exclusion, expedience, and obedience to material forces what we really want to think of as the sources of our origins.

In Max K. Hecht (ed.)
Evolutionary Biology at the Crossroads
Commentaries (p. 175)

Simpson, George Gaylord
When bright young biologists speak of genetics without genes and wise old biologists of life without organisms it is evident that something peculiar is going on in the science of biology, so peculiar that "crisis" is not too strong a word. I would diagnose this as combining monomania and schizophrenia.

Biology and Man
Chapter 1 (p. 3)

Steinbeck, John
We sat on crates of oranges and thought what good men most biologists are, the tenors of the scientific world—temperamental, moody, lecherous, loud laughing and healthy... Your true biologist will sing you a song as loud and off-key as will a blacksmith, for he knows that morals are too often diagnostic of prostatitis and stomach ulcers. Sometimes he may proliferate a little too much in all directions, but he is as easy to kill as any other organism, and meanwhile he is very good company, and at least he does confuse a low hormone productivity with moral ethics.

The Log from the Sea of Cortez
Chapter 4 (p. 28, 28–9)

Stockbridge, Frank B.
"A little bit of this, a little more of that, a pinch of something else, boil blank minutes, and set aside in the same vessel"—thus might read the biologists' formula for creating life...

Cosmopolitan
Creating Life in the Laboratory (p. 775)
May 1912

Unknown
A group of goose biologists were meeting to brainstorm about the migration tactics of Canada geese. They were particularly interested in applying for a $100,000 Federal grant to investigate the "V" formation of goose flight. It had been observed that one side of the "V" is always longer

than the other side. This group would put together a research proposal to apply for the $100,000 grant and hopefully find out why this happens.

To start off the discussion, Todd, the Consulting Firm Biologist, stands up and says in typical consultant fashion, "I say we ask for $200,000, and attempt to model the wind drag coefficients. We can have our geologists record and map the ground topography and then our staff meteorologists can predict potential updraft currents. Our internal CAD department can then produce 3D drawings of the predicted wing tip vortices. Then, after several years of study, our in-house publications department could produce a nice thick report full of charts and graphs."

The Senior Research Biologist, a professor at the local university, cleared his throat and responded, "No, no! That's not it at all. We only need $150,000. We can train a group of domesticated geese to fly in formations of equal length and then compare their relative fitness to wild geese. We can then publish the results in the *Journal of Wildlife Management*.

About then, the hardworking field biologist stands up and begins walking for the door. "Where are you going?" the group asks. "I'm leaving" he replies, "I've heard enough. No one has to give me $100,000 to find out that the reason one side of the "V" is longer is simply because there are more geese on that side!"

Source unknown

Vogel, Steven
With the ratification of long tradition, the biologist goes forth, thermometer in hand, and measures the effects of temperature on every parameter of life. Lack of sophistication poses no barrier; heat storage and exchange may be ignored or Arrhenius abused; but temperature is, after time, our favorite abscissa. One doesn't have to be a card-carrying thermodynamicist to wield a thermometer.

Life in Moving Fluids
Chapter 1 (p. 1)

Wilson, Edward O.
The role of science, like that of art, is to blend proximate imagery with more distant meaning, the parts we already understand with those given as new into larger patterns that are coherent enough to be acceptable as truth. Biologists know this relation by intuition during the course of fieldwork, as they struggle to make order out of the infinitely varying patterns of nature.

In Search of Nature
The Bird of Paradise: The Hunter and the Poet (p. 129)

BIOLOGY

Capra, Fritjof

The exploration of the atom has forced physicists to revise their basic concepts about the nature of physical reality in a radical way. The result of the revision is a coherent dynamic theory, quantum mechanics, which transcends the principal concepts of Cartesian–Newtonian science. In biology, on the other hand, the exploration of the gene has not led to a comparable revision of basic concepts, nor has it resulted in a universal dynamic theory.

The Turning Point
Chapter 4 (p. 121)

Carson, Rachel

The "control of nature" is a phrase conceived in arrogance, born of the Neanderthal age of biology and philosophy, when it was supposed that nature exists for the convenience of man.

Silent Spring
Chapter 17 (p. 297)

Chargaff, Erwin

In the old times, the knowledge of biology was perhaps similar to what could be made out in a very large, very dark house. Many objects could be more felt than seen with equal dimness, once the eyes got used to the darkness; and scientists were conscious of the limiting conditions under which they worked. In our time, however, a few very powerful and very narrow beams of light have been thrown into a few corners of this dark house, and several things can be seen in clarity and illumination that almost distort their significance. But at the same time we have lost our dark-adaptation; and since we all have a tendency to follow the light, we have moved into these cozy corners, to the detriment of the rest, which still is, by far, the major part of nature. In pointing this out one runs the risk of being accused of trying to spread the darkness.

Essays on Nucleic Acids
Chapter 3 (pp. 39–40)

Cohen, Joel
Physics-envy is the curse of biology.

Science
Mathematics as Metaphor (p. 675)
Volume 172, 14 May 1971

Crick, Francis Harry Compton
The ultimate aim of the modern movement in biology is in fact to explain all biology in terms of physics and chemistry.

Of Molecules and Men
The Nature of Vitalism (p. 10)

Dawkins, Richard
Biology is the study of the complex things in the Universe. Physics is the study of the simple ones.

New Scientist
The Necessity of Darwinism (p. 130)
Volume 94, Number 1301, 15 April 1982

Delbrück, Max
Biology is a very interesting field to enter for anyone, by the vastness of its structure and the extraordinary variety of strange facts it has collected, but to the physicist it is also a depressing subject, because, insofar as physical explanations of seemingly physical phenomena go, like excitation, or chromosome movements, or replication, the analysis seems to have stalled around in a semi descriptive manner without noticeably progressing towards a radical physical explanation. He may be told that the only access of atomic physics to biology is through biochemistry. Listening to the story of modern biochemistry he might become persuaded that the cell is a sack full of enzymes acting on substrates converting them through various intermediate stages either into cell substance or into waste products... It looks sane until paradoxes crop up and come into sharper focus. In biology we are not yet at the point where we are presented with clear paradoxes and this will not happen until the analysis of the behavior of living cells has been carried into far greater detail.

Transactions of the Connecticut Academy of Sciences
A Physicist Looks at Biology (p. 191)
Volume 38, 1949

Dobzhansky, Theodosius
Seen in the light of evolution, biology is, perhaps, intellectually the most satisfying and inspiring science. Without that light it becomes a pile of sundry facts—some of them interesting or curious but making no meaningful picture as a whole.

In J. Peter Zetterberg (ed.)
Evolution versus Creationism (p. 3)

Driesch, H.
The analysis of the Aristotelian theory of life must therefore be one of the corner-stones of any historical works on biology.

The History & Theory of Vitalism
Chapter I (p. 11)

Emmeche, Claus
Biology belongs to one of the surprising sciences, where each rule must always be supplemented with several exceptions (except this rule, of course).

The Garden in the Machine
Chapter 6 (p. 144)

Fauset, Jessie Redmon
Biology transcends society!

The Chinaberry Tree
Chapter XIX (p. 121)

Freud, Sigmund
Biology is truly a land of unlimited possibilities; we may have the most surprising revelations to expect from it, and cannot conjecture what answers it will offer in some decades to the questions we have put to it. Perhaps they may be such as to overthrow the whole artificial structure of hypotheses.

Beyond the Pleasure Principle
Chapter VI (p. 78)

Goodwin, Brian Carey
The discovery of appropriate variables for biology is itself an act of creation.

In C.H. Waddington (ed.)
Towards a Theoretical Biology
Volume 2
Appendix notes on the second symposium (p. 337)

Gore, Rick
If anything illustrates what has happened in biology, it is this profound new ability to take the very stuff of life out of a cell, to isolate it in a test tube, to dissect it, and to probe the deep mysteries borne in its fragments.

National Geographic
The Awesome Worlds Within a Cell (p. 355)
Volume 150, Number 3, September 1976

Grassé, Pierre P.
Biology, despite the brilliance of its appearance, stammers in the presence of the essentials. We know neither all the properties of living matter, nor all of its astonishing possibilities.

In Joseph Wood Krutch
The Great Chain of Life
Chapter 11 (p. 192)

Grobstein, Clifford
From studies aimed at molecules, cells, organisms and populations will come a global conception of earth's biotic film, and from this a projection of this concept to the universe at large. Confidence that we shall achieve this conception also characterizes today's biology. Excitement, confidence, and expectation are in the air, as though all that we now know and say of life is but a prologue.

The Strategy of Life
Chapter 1 (p. 6)

Haldane, J.B.S.
If physics and biology one day meet, and one of the two is swallowed up, that one will not be biology.

In J. Needham
Time: The Refreshing River
A Biologist's View of Whitehead's Philosophy (p. 204)

Hoyle, Fred
I wouldn't go into biology if I were starting again now. In twenty years' time it is the biologists who will be working behind barbed wire.

In G. Rattray Taylor
The Biological Time Bomb
Chapter I (p. 17)

Huxley, Aldous

Solved by standard Gammas, unvarying Deltas, uniform Epsilons. Millions of identical twins. The principle of mass production at last applied to biology.

Brave New World
Chapter 1 (pp. 6–7)

Huxley, Thomas H.

In the first place it is said—and I take this point first, because the imputation is too frequently admitted by Physiologists themselves—that Biology differs from the Physico-chemical and Mathematical sciences in being "inexact".

Lay Sermons, Addresses, and Reviews
On the Educational Value of the Natural History Sciences (pp. 78–9)

Kauffman, Stuart

If biologists have ignored self-organization, it is not because self-ordering is not pervasive and profound. It is because we biologists have yet to understand how to think about systems governed simultaneously by two sources of order. Yet who seeing the snowflake, who seeing simple lipid molecules cast adrift in water forming themselves into cell-like hollow lipid vesicles, who seeing the potential for the crystallization of life in swarms of reacting molecules, who seeing the stunning order for free in networks linking tens upon tens of thousands of variables, can fail to entertain a central thought: if ever we are to attain a final theory in biology, we will surely, surely have to understand the commingling of self-organization and selection. We will have to see that we are the natural expressions of a deeper order. Ultimately, we will discover in our creation myth that we are expected after all.

At Home in the Universe
Chapter 5 (p. 112)

Lamarck, Jean Baptiste Pierre Antoine

A sound *Physics of the Earth* should include all the primary considerations of the earth's atmosphere, of the characteristics and continual changes of the earth's external crust, and finally of the origin and development of living organisms. These considerations naturally divide the physics of the earth into three essential parts, the first being a theory of the atmosphere, or *Meteorology*, the second a theory of the earth's external crust, or *Hydrogeology*, and the third a theory of living organisms, or *Biology*.

Hydrogeology
Forward (p. 18)

Lorenz, Konrad

There are no good biologists whose vocation was not born of deep joy in the beauties of living nature.

In Jean Rostand
Humanly Possible
A Biologist's Mail (p. 20)

Monod, Jacques

Biology occupies a position among the sciences at once marginal and central. Marginal because—the living world constituting but a tiny and very "special" part of the universe—it does not seem likely that the study of living beings will ever uncover general laws applicable outside the biosphere. But if the ultimate aim of the whole of science is indeed, as I believe, to clarify man's relationship to the universe, then biology must be accorded a central position since of all disciplines it is the one that endeavors to go most directly to the heart of the problems that must be resolved before that of 'human nature' can be framed in other than metaphysical terms.

Chance and Necessity
Preface (p. xi)

Needham, James G.

It is a monstrous abuse of the science of biology to teach it only in the laboratory...Life belongs in the fields, in the ponds, on the mountains and by the seashore.

In Allen H. Benton and William E. Werner
Field Biology and Ecology (p. 3)

Osler, Sir William

Biology touches the problems of life at every point, and may claim, as no other science, completeness of view and a comprehensiveness which pertains to it alone. To all those whose daily work lies in her manifestations the value of a deep insight into her relations cannot be overestimated. The study of biology trains the mind in accurate methods of observation and correct methods of reasoning, and gives to a man clearer points of view, and an attitude of mind serviceable in the working-day-world than that given by other sciences, or even by the humanities.

Aequanimitas
The Leaven of Science (pp. 91–2)

Roberts, Catherine
The driving force of biology and medical science is not unalloyed idealism but a complex of factors including prestige, publication, professional advancement, grants and business interests.

Perspectives in Biology and Medicine
The Use of Animals in Medical Research—
Some Ethical Considerations (p. 116, fn 4)
Volume VIII, Number 1, Autumn 1964

Root, R.K.
I can hear my good friend, the Professor of Biology, rather impatiently reporting that his science asks assent only to what it can demonstrate. "Come with me to my laboratory, and I will give you proofs..." But how am I, quite untrained in his science, to weigh his arguments or interpret what his microscopes may show?

The Atlantic Monthly
The Age of Faith (p. 114)
Volume CX, July 1912

Rostand, Jean
[Biology] is the least self-centered, the least narcissistic of the sciences— the one that, by taking us out of ourselves, leads us to re-establish a link with nature and to shake ourselves free from our spiritual isolation.

Can Man be Modified?
Victories and the Hopes of Biology (p. 31)

Simpson, George Gaylord
Biology, then, is the science that stands at the center of all science. It is the science most directly aimed at science's major goal and most definitive of that goal. And it is here, in the field where all the principles of all the sciences are embodied, that science can truly become unified.

This View of Life: The World of an Evolutionist
Chapter 5 (p. 107)

Simpson, George Gaylord
Pittendrigh, Colin S.
Tiffany, Lewis
We believe that there is a unified science of life, a general biology that is distinct from a shotgun marriage of botany and zoology, or any others of the special life sciences. We believe that this science has a body of established and working principles. We believe that literally nothing on

earth is more important to a rational living than basic acquaintance with those principles.

Life: An Introduction to Biology
2nd Edition
Preface from 1st Edition (p. v)

Standen, Anthony
In its central content, biology is not accurate thinking, but accurate observation and imaginative thinking, with great sweeping generalizations.

Science is a Sacred Cow
Chapter IV (pp. 99–100)

Sullivan, J.W.N.
It is possible, nevertheless, that our outlook on the physical universe will again undergo a profound change. This change will come about through the development of biology. If biology finds it absolutely necessary, for the description of living things, to develop new concepts of its own, then the present outlook on "inorganic nature" will also be profoundly affected... The notions of physics will have to be enriched, and this enrichment will come from biology.

The Limitations of Science
Towards the Future (pp. 188, 189)

Unknown
Biology is the only science in which multiplication means the same thing as division.

Source unknown

Biology is really Chemistry, Chemistry is really Physics, Physics is really Mathematics, and Mathematics is really Philosophy.

Source unknown

Weaver, W.
The century of biology upon which we are now well embarked is no matter of trivialities. It is a movement of really heroic dimensions, one of the great episodes in man's intellectual history. The scientists who are carrying the movement forward talk in terms of nucleoproteins, of ultra-centrifuges, of biochemical genetics, of electrophoresis, of the electron microscope, or molecular morphology, of radioactive isotopes. But do not be fooled into thinking this is more gadgetry. This is the dependable way to seek a solution of the cancer and polio problems, the problem of rheumatism and of the heart. This is the knowledge on which we

must base our solution of the population and food problems. This is the understanding of life.

In R.B. Fosdick
The Rockefeller Foundation (p. 166)
Letter to H.M.H. Carson
17 June 1949

Whitehead, Alfred North

The living cell is to biology what the electron and the proton are to physics.

Science and the Modern World
The Nineteenth Century (p. 146)

Science is taking on a new aspect that is neither purely physical nor purely biological. It is becoming the study of the larger organisms; whereas physics is the study of the smaller organisms.

Science and the Modern World
The Nineteenth Century (p. 150)

Accordingly, biology apes the manners of physics. It is orthodox to hold, that there is nothing in biology but what is physical mechanism under somewhat complex circumstances.

Science and the Modern World
The Nineteenth Century (p. 150)

Unfortunately in this book of nature the biologists fare badly. Every expression of life takes time. Nothing that is characteristic of life can manifest itself at an instant. Murder is a prerequisite for the absorption of biology into physics as expressed in these traditional concepts.

Aristotelian Society
Supplementary
Volume II (p. 45)
Time, Space and Material

Woodger, Joseph Henry

Biology is being forced in spite of itself to become biological.

In Herbert J. Muller
Science and Criticism
Chapter V (p. 110)

BIRDS

Atkinson, Brooks

Nothing wholly admirable ever happens in this country except the migration of birds.

Once Around the Sun
March 23 (p. 80)

Chapman, Frank M.

...birds will appeal most strongly to us through their songs. When your ears are attuned to the music of birds, your world will be transformed. Birds' songs are the most eloquent of Nature's voices...

Bird-Life
Chapter I (p. 11)

Cornwall, Barry

Come, all ye feathery people of mid-air,
Who sleep 'midst rocks, or on the mountain summits
Lie down with the wild winds; and ye who build
Your homes amidst green leaves by grottoes cool;
And ye who on the flat sands hoard your eggs
For suns to ripen, come!

The Poetical Works of Milman, Bowles, Wilson, and Barry Cornwall
An Invocation to Birds

Darwin, Charles

We behold the face of nature bright with gladness, we often see the superabundance of food; we do not see or we forget, that the birds which are idly singing round us mostly live on insects or seeds, and are thus constantly destroying life; or we forget how largely these songsters, or their eggs, or their nestlings, are destroyed by birds and beasts of prey; we do not always bear in mind, that, though food may be now superabundant, it is not so at all seasons of each recurring year.

The Origin of Species
Chapter III (p. 32)

Emerson, Ralph Waldo

The bird is not in its ounces and inches, but in its relations to Nature; and the skin or skeleton you show me, is no more a heron, than a heap of ashes or a bottle of gases into which his body has been reduced, is Dante or Washington.

The Conduct of Life
Beauty (pp. 247–8)

Huxley, Julian

Birds in general are stupid, in the sense of being little able to meet unforeseen emergencies; but their lives are often emotional, and their emotions are richly and finely expressed.

Essays of a Biologist
An Essay on Bird-Mind (p. 109)

Klee, Paul

The birds are to be envied:
They avoid
Thinking about the trees and the roots.
Agile, self contented, all day long they swing
And sing, perched on ultimate end.

The Inward Vision (Cover page)

Lawrence, D.H.

I never saw a wild thing
Sorry for itself.
A small bird will drop frozen dead
From a bough
Without ever having felt sorry for itself.

The Complete Poems of D.H. Lawrence
Volume I
Self Pity

Longfellow, Henry Wadsworth

You call them thieves and pillagers; but know,
They are the winged wardens of your farms,
Who from the cornfields drive the insidious foe,
And from your harvests keep a hundred harms;...

The Complete Writings of Henry Wadsworth Longfellow
Volume IV
The Poet's Tale
Birds of Killingworth
Stanza 19

FUNNY THING IS —
I DON'T FEEL AT ALL
SORRY FOR MYSELF..!

Lynd, Robert
There is nothing in which the birds differ more from man than the way in which they can build and yet leave a landscape as it was before.

The Blue Lion
The Nuthatch (p. 29)

Mansfield, Katherine
It is astonishing how violently a big branch shakes when a silly little bird has left it. I expect the bird knows it and feels immensely arrogant.

In J. Middleton Murry (ed.)
Journal of Katherine Mansfield
1917
August 21
Alors, je pars (p. 70)

McArthur, Peter
The robins, killdeere, red-winged blackbirds and grackles come back with the warm wave. This means that the great university of nature is about to open for its spring and summer terms.

The Best of Peter McArthur
Nature's University (pp. 169–70)

Pallister, William
Of the BIRDS, thirteen thousands of species are named;
This is the first life with warm blood! We could not know all
And quite truly one need not feel greatly ashamed

If some few of the rare names are hard to recall,
But the birds are so lovely, I wish that I knew
All about all of them, and I'm sure so do you.

Poems of Science
Beginnings
Animal Life (p. 141)

Unknown
Der spring is sprung
Der grass is riz
I wonder where dem boidies is?

Der little boids is on der wing,
Ain't dat absoid?
Der little wings is on der boid!

In Arnold Silcock
Verse and Worse
The Budding Bronx (p. 37)

Whitman, Walt
You must not know too much, or be too precise or scientific about birds
and trees and flowers and watercraft; a certain free margin, and even
vagueness—perhaps ignorance, credulity—helps your enjoyment of these
things.

Specimen Days
Birds—And A Caution (p. 112)

ALBATROSS

Coleridge, Samuel T.
"God save thee, ancient Mariner,
From the fiends that plague thee thus!—
Why look'st thou so?"—"With my cross-bow
I shot the Albatross."

In Max J. Herzberg (ed.)
Narrative Poems
The Ancient Mariner
Part I, Stanza 20

Leland, Charles G.
Great albatross!—the meanest birds
Spring up and flit away,
While thou must toil to gain a flight,

And spread those pinions grey;...

The Music-Lesson of Confucius
'Perseverando'
Stanza 3

BIRD OF PARADISE

Colum, Padraic
With sapphire for her crown,
And with the Libyan wine
For lustre of her eyes;
With azure for her feet
(It is her henna stain);
Then iris for her vest,
Rose, ebony, and flame,
She lives a thing enthralled,
In forests that are old,
As old as is the Moon.

Poems
Bird of Paradise

Moore, Thomas
Those golden birds that, in the spice-time, drop
About the gardens, drunk with that sweet food
Whose scent hath lur'd them o'er the summer flood
And those that under Araby's soft sun
Build their high nests of budding cinnamon.

The Poetical Works of Thomas Moore
Lalla Rookh
The Veiled Prophet of Khorassan (p. 48)

BLACKBIRD

Moir, D.M.
The birds have ceased their songs,
All save the blackbird, that from yon tall ash,
'Mid Pinkie's greenery, from his mellow throat,
In adoration of the setting sun,
Chants forth his evening hymn.

The Poetical Works of David Macbeth Moir
An Evening Sketch

Twain, Mark
The blackbird is a perfect gentleman, in deportment and attire, and is not noisy, I believe, except when holding religious services and political conventions in a tree...

Following the Equator
Volume II
Chapter II (p. 32)

BLUEBIRD

Longfellow, Henry Wadsworth
In the thickets and the meadows
Piped the bluebird, the Owaissa...

The Complete Writings of Henry Wadsworth Longfellow
Volume II
Hiawatha
Part XXI

BOBOLINK

Cranch, C.P.
One day in the bluest of summer weather,
Sketching under a whispering oak,
I heard five bobolinks laughing together,
Over some ornithological joke.

Collected Poems of Christopher Pearse Cranch
Bird Language
Stanza I

CANARY

Mulock, Dinah Maria
Sing away, ay, sing away,
Merry little bird,
Always gayest of the gay,
Though a woodland roundelay
You ne'er sung nor heard;
Though your life from youth to age
Passes in a narrow cage.

Miss Mulock's Poems
The Canary in His Cage

Nash, Ogden
The song of canaries
Never varies,
And when they're moulting
They're pretty revolting.

Verses from 1929 On
The Canary

CAPE MAY WARBLER

Halle, Louis J.
When I see men able to pass by such a shining and miraculous thing as this Cape May warbler, the very distillate of life, and then marvel at the internal-combustion engine, I think we had better make ourselves ready for another Flood.

Spring in Washington
Chapter II (p. 74)

CROW

Gay, John
To shoot at crows is powder flung away.

The Poetical Works of John Gay
Volume I
Epistle to the Right Honourable Paul Methuen, Esq., l. 96

Longfellow, Henry Wadsworth
Even the blackest of them all, the crow,
Renders good service as your man-at-arms,
Crushing the beetle in his coat of mail,
And crying havoc on the slug and snail.

The Complete Writings of Henry Wadsworth Longfellow
Volume IV
The Poet's Tale
Birds of Killingworth
Stanza 19

CUCKOO

Shakespeare, William
...the cuckoo builds not for himself...

Anthony and Cleopatra
Act II, Scene VI, l. 28

DODO

Cuppy, Will
The Dodo never had a chance. He seems to have been invented for the sole purpose of becoming extinct and that was all he was good for.

How to Become Extinct
The Dodo (p. 102)

DOVE

Browning, Elizabeth Barrett
And there my little doves did sit
With feathers softly brown,
And glittering eyes that showed their right
To general Nature's deep delight.

The Complete Poetical Works of Elizabeth Barrett Browning
My Doves
Stanza 2

Shakespeare, William
The dove and very blessed spirit of peace...

The Second Part of King Henry the Fourth
Act IV, Scene I, l. 46

DUCK

Adams, Douglas
...even the sceptical mind must be prepared to accept the unacceptable when there is no alternative. If it looks like a duck, and quacks like a duck, we have at least to consider the possibility that we have a small aquatic bird of the family *Anatidae* on our hands.

Dirk Gently's Holistic Detective Agency
Chapter 30 (p. 216)

EAGLE

Tennyson, Alfred
He clasps the crag with hooked hands,
Close to the sun in lonely lands;
Ring'd with the azure world, he stands.

The wrinkled sea beneath him crawls;
He watches from his mountain walls,
And like a thunderbolt he falls.

The Complete Poetical Works of Tennyson
The Eagle

EMU

Prelutsky, Jack
Do not approach an emu,
The bird does not esteem you.
It wields a quick and wicked kick
That's guaranteed to cream you.

A Pizza the Size of the Sun
Do Not Approach an Emu

FALCON

Lowell, Maria White
I know a falcon swift and peerless
As e'er was cradled in the pine:
No bird had ever eye so fearless,
Or wing so strong as this of mine.

The Falcon

Shakespeare, William
My falcon now is sharp and passing empty;
And till she stoop, she must not be full-gorged,
For then she never looks upon her lure.

The Taming of the Shrew
Act IV, Scene I, l. 193–5

GOLDFINCH

Cowper, William
Two goldfinches, whose sprightly song
Had been their mutual solace long,
Lived happy prisoners there.

The Poetical Works of William Cowper
Faithful Bird

Dryden, John
A goldfinch there I saw, with gawdy pride

Of painted plumes, that hopped from side to side,
Still pecking as she pass'd; and still she drew
The sweets from every flower, and suck'd the dew:
Sufficed at length, she warbled in her throat,
And turned her voice to many a merry note...

The Poetical Works of John Dryden
Tales From Chaucer
The Flower and the Leaf, l. 106–11

GOOSE

Shakespeare, William
As wild geese that the creeping fowler eye,
Or russet-pated choughs, many in sort,
Rising and cawing at the gun's report,
Sever themselves, and madly sweep the sky.

A Midsummer-Night's Dream
Act III, Scene II, l. 20–3

Young, Roland
The plural of goose is geese,
But the plural of moose ain't meese,
And the plural of noose ain't neese,
But the plural of goose—*is* geese.

Not for Children
The Goose

GRACKLE

Nash, Ogden
The grackle's voice is less than mellow,
His heart is black, his eye is yellow,
He bullies more attractive birds
With hoodlum deeds and vulgar words,
And should a human interfere,
Attacks that human in the rear.
O cannot help but deem the grackle
An ornithological debacle.

Verses from 1929 On
The Grackle

GULL, SEA

Shakespeare, William
And being fed by us you used us so
As that ungentle gull, the cuckoo's bird,
Useth the sparrow…

The First Part of King Henry the Fourth
Act V, Scene I, l. 59–61

HAWK

Shakespeare, William
When I bestride him, I soar, I am a hawk…

The Life of King Henry the Fifth
Act III, Scene VII, l. 14

Tennyson, Alfred
The wild hawk stood with the down on his beak,
And stared with his foot on the prey.

The Complete Poetical Works of Tennyson
The Poet's Song, l. 11–12

HUMMING-BIRD

Pallister, William
A flashing, dashing, rainbow-streak,
The whir of wondrous wings;
We hold our breath, we must not speak,
Such shy, such splendid things!

Poems of Science
De Ipsa Natura
Humming-Birds (p. 222)

Riley, James Whitcomb
And the humming-bird that hung
Like a jewel up among
The tilted honeysuckle-horns,
They mesmerized and swung
In the palpitating air,
Drowsed with odors strange and rare,
And, with whispered laughter, slipped away,

And left him hanging there.

The Complete Works of James Whitcomb Riley
Volume IV
The South Wind and the Sun
Stanza 8

Tabb, John Banister
A flash of harmless lightning,
A mist of rainbow dyes,
The burnished sunbeams brightening,
From flower to flower he flies:...

The Poetry of Father Tabb
Birds
The Humming-Bird

JAY

Shakespeare, William
What is the jay more precious than the lark,
Because his feathers are more beautiful.

The Taming of the Shrew
Act IV, Scene III, l. 177–8

LARK

Browning, Elizabeth Barrett
The music soars within the little lark,
And the lark soars.

The Complete Poetical Works of Elizabeth Barrett Browning
Aurora Leigh
Book III, l. 155–6

Rossetti, Christina G.
The sunrise wakes the lark to sing.

The Complete Poems of Christina Rossetti
Volume I
Poems Added in 1875
Bird Raptures (p. 210)

LINNET

Wordsworth, William
Hail to thee, far above the rest

In joy of voice and pinion!
Thou, linnet! in thy green array,
Presiding spirit here to-day,
Dost lead the revels of the May;
And this is thy dominion.

The Complete Poetical Works of William Wordsworth
The Green Linnet
Stanza II

LOON

Lawrence, Jerome
Lee, Robert E.
Anytime you hear a man called "loony," just remember that's a great compliment to the man and a great disrespect to the loon. A loon doesn't wage war, his government is perfect, being nonexistent. He is the world's best fisherman and completely in control of his senses, thank you.

The Night Thoreau Spent in Jail
Act 1 (p. 12)

MARTLET

Shakespeare, William
. . . the martlet
Builds in the weather on the outward wall,
Even in the force and road of casuality.

The Merchant of Venice
Act II, Scene IX, l. 28–30

MOCKING-BIRD

Longfellow, Henry Wadsworth
Then from the neighboring thicket the mocking-bird, wildest of singers,
Swinging aloft on a willow spray that hung o'er the water,
Shook from his little throat such floods of delirious music,
That the whole air and the woods and the waves seemed silent to listen.

The Complete Writings of Henry Wadsworth Longfellow
Volume II
Evangeline
Part II, Stanza II (pp. 75–6)

NIGHTINGALE

Sappho
The nightingale is the harbinger of Spring and her voice is desire.

Poems and Fragments
Fragment 114

OSTRICH

Nash, Ogden
The ostrich roams the great Sahara.
Its mouth is wide, its neck is narra.
It has such long and lofty legs,
I'm glad it sits to lay its eggs.

Verses from 1929 On
The Ostrich

OWL

Borland, Hal
The owl, that bird of onomatopoetic name, is a repetitious question wrapped in feathery insulation especially for Winter delivery.

Sundial of the Seasons
Questioner
December 27 (p. 271)

Shakespeare, William
The clamorous owl that nightly hoots and wonders
At our quaint spirits.

A Midsummer-Night's Dream
Act II, Scene II, l. 6–7

PARROT

Prelutsky, Jack
The parrots, garbed in gaudy dress,
With almost nothing to express,
Delight in spouting empty words...
They are extremely verbal birds.

A Pizza the Size of the Sun
The Parrots

PARTRIDGE

Spenser, Edmund
Like as a feareful partridge, that is fled
From the sharpe hauke which her attacked neare,
And falls to ground to seeke for succor theare,
Whereas the hungry spaniells she does spye,
With greedy jaws her ready for to teare.

The Complete Poetical Works of Edmund Spenser
Faerie Queene
Book III, Canto VIII, Stanza 33

PEACOCK

Leland, Charles G.
To Paradise, the Arabs say,
Satan could never find the way
Until the peacock led him in…

The Music-Lesson of Confucius
The Peacock
Stanza 2

Shakespeare, William
Why, he stalks up and down like a peacock,—a stride and a stand…

Troilus and Cressida
Act III, Scene III, l. 251

PELICAN

Merritt, Dixon L.
A wonderful bird is the pelican!
His bill will hold more than his belican.
He can take in his beak
Food enough for a week
But I'm darned if I see how the helican.

The Pelican
Source unknown

Montgomery, James
Bird of the wilderness, what is thy name?
—The pelican!—go, take the trump of fame,
And if thou give the honour due to me,

The world may talk a little more of thee.

Poetical Works of James Montgomery
Volume II
Birds

PENGUIN

Herford, Oliver
The Pen-guin sits up-on the shore
And loves the lit-tle fish to bore;
He has one en-er-vat-ing joke
That would a very Saint pro-voke:
"The *Pen*-guin's might-I-er than the *Sword*-fish"
He tell this dai-ly to the bored fish,
Un-til they are so weak, they float
With-out re-sis-tance down his throat.

A Child's Primer of Natural History
A Penguin

Young, Roland
The little penguins look alike
Even as Ike resembles Mike.
They are so gentle and so nice
God keeps these little birds on ice.

Not for Children
The Penguin

PHEASANT

Pope, Alexander
See! from the brake the whirring pheasant springs,
And mounts exhaulting on triumphant wings:
Short is his joy; he feels the fiery wound,
Flutters in blood, and panting beats the ground.

Alexander Pope's Collected Poems
Windsor Forest
l. 111–14

PIGEON

Willis, Nathaniel Parker
On the cross-beam under the Old South bell

The nest of a pigeon is builded well.
In summer and winter that bird is there,
Out and in with the morning air...

Poems of Nathaniel Parker Willis
The Belfry Pigeon

QUAIL

Longfellow, Henry Wadsworth
The song-birds leave us at the summer's close,
Only the empty nests are left behind,
And pipings of the quail among the sheaves.

The Complete Writings of Henry Wadsworth Longfellow
Volume III
The Harvest Moon

RAVEN

Poe, Edgar Allan
And the raven, never flitting, still is sitting, still is sitting
On the pallid bust of Pallas just above my chamber door;
And his eyes have all the seeming of a demon's that is dreaming,
And the lamplight o'er him streaming throws his shadow on the floor;
And my soul from out that shadow that lies floating on the floor
Shall be lifted—nevermore!

The Raven and Other Poems
The Raven
Stanza 18

ROBIN

Lowell, Maria White
Who killed Cock Robin?
"I," said the Sparrow,
"With my bow and arrow,
I killed Cock Robin."

Nursery Rhyme

Blake, William
A Robin Red breast in a Cage
Puts all Heaven in a Rage.

The Complete Poetry and Prose of William Blake
Auguries of Innocence, l. 5–6

ROOK

Tennyson, Alfred
The building rook'll caw from the windy tall elm-tree...

The Complete Poetical Works of Tennyson
The May Queen
New Year's Eve, Stanza 5

SAND-PIPER

Thaxter, Celia
Across the narrow beach we flit,
One little sandpiper and I;
And fast I gather, bit by bit,
The scattered driftwood, bleached and dry.
The wild waves reach their hands for it,
The wild wind raves, the tide runs high,
As up and down the beach we flit,
One little sandpiper and I.

The Poems of Celia Thaxter
The Sand-Piper

SEA-MEW

Browning, Elizabeth Barrett
How joyously the young sea-mew
Lay dreaming on the waters blue,
Whereon our little bark had thrown
A little shade, the only one,
But shadows ever man pursue.

The Complete Poetical Works of Elizabeth Barrett Browning
The Sea-Mew
Stanza I

Garstang, Walter
Bold Sea-mew—you whose soaring flight
Inspires my envious Muse—
Pray, with this compliment polite
My liberty excuse.

Larval Forms
To A Herring Gull
Stanza 1 (p. 72)

SEDGE-BIRD

Clare, John

Fixed in a white-thorn bush, its summer guest,
So low, e'en grass o'er-topped its tallest twig,
A sedge-bird built its little benty nest,
Close by the meadow pool and wooden brig.

The Rural Muse
Poems
The Sedge-Bird's Nest

SPARROW

Longfellow, Henry Wadsworth

The sparrows chirped as if they still were proud
Their race in Holy Writ should mentioned be.

The Complete Writings of Henry Wadsworth Longfellow
Volume IV
The Poet's Tale
The Birds of Killingworth
Stanza 12

Shakespeare, William

The hedge-sparrow fed the cuckoo so long,
That it had its head bit off by its young.

King Lear
Act I, Scene IV, l. 235–6

SWALLOW

Longfellow, Henry Wadsworth

The swallow is come!
The swallow is come!
O, fair are the seasons, and light
Are the days that she brings,
With her dusky wings,
And her bosom snowy white.

Hyperion
Book II, Chapter I

Thomson, James
The swallow sweeps
The slimy pool, to build his hanging house.

The Seasons
Spring, l. 651

Tennyson, Alfred
...nature's licensed vagabond, the swallow...

The Complete Poetical Works of Tennyson
Queen Mary
Act V, Scene I

SWAN

Beston, Henry
...I chanced to look up a moment at the southern sky, and there for the first and still the only time in my life, I saw a flight of swans. The birds were passing along the coast well out to sea; they were flying almost cloud high and traveling very fast, and their course was as direct as an arrow's from a bow. Glorious white birds in the blue October Heights over the solemn unrest of ocean—their passing was more than music, and from their wings descended the old loveliness of earth which both affirms and heals.

The Outermost House
Autumn, Ocean, and Birds (p. 37)

Thomson, James
The stately-sailing swan
Gives out his snowy plumage to the gale;
And, arching proud his neck, with oary feet
Bears forward fierce, and guards his osier isle,
Protective of his young.

The Seasons
Spring, l. 775

THROSTLE

Wordsworth, William
And hark! How blithe the throstle sings!
He, too, is no mean preacher:
Come forth into the light of things,

Let Nature be your teacher.

The Complete Poetical Works of William Wordsworth
The Tables Turned
Stanza IV

THRUSH

Tennyson, Alfred
When rosy plumelets tuft the larch,
And rarely pipes the mounted thrush...

The Complete Poetical Works of Tennyson
In Memoriam
Part XCI

Hardy, Thomas
At once a voice arose among
The bleak twigs overhead
In a full-hearted evensong
Of joy illimited;
An aged thrush, frail, gaunt, and small,
In blast-beruffled plume,
Had chosen thus to fling his soul
Upon the growing gloom.

Collected Poems of Thomas Hardy
The Darkling Thrush
Verse 3

TOUCAN

Wood, Robert William
Very few can
Tell the Toucan
From the Pecan—
Here's a new plan:
To take the Toucan from the Tree,
Requires im-mense a-gil-i-tee,
While anyone can pick with ease
The Pecans from the Pecan trees.
It's such an easy thing to do,
That even the Toucan he can too.

How to Tell the Birds from the Flowers and Other Woodcuts
The Pecan. The Toucan. (p. 11)

Nash, Ogden
The toucan's profile is prognathous,
Its person is a thing of bathos.
If even I can tell a toucan
I'm reasonably sure that you can.

Verses from 1929 On
The Toucan

VULTURE

Belloc, Hilaire
The Vulture eats between his meals,
And that's the reason why
He very, very rarely feels
As well as you and I.

Complete Verse
The Vulture

Montgomery, James
Abdominal harpies, spare the dead.
—We only clear the field which man has spread;
On which should Heaven its hottest vengeance rain?
You slay the living, we but strip the dead.

Poetical Works of James Montgomery
Volume II
Birds

WHITE-THROAT

Clare, John
The happy white-throat on the swinging bough,
Rocked by the impulse of the gadding wind
That ushers in the showers of April, now
Carols right joyously; and now reclined
Crouching, she clings close to her moving seat,
To keep her hold.

The Rural Muse
Poems
The Happy Bird

WREN

Wordsworth, William
Among the dwellings framed by birds
In fields or forests with nice care,
Is none that with the little wren's
In snugness may compare.

The Complete Poetical Works of William Wordsworth
A Wren's Nest
Stanza I

BOOK

de Beer, Sir G.
There is a small number of great books which have changed the face of the earth. Such as the Bible, they have exerted their effects universally even if their tenants are not everywhere accepted. Others, like Newton's *Principia*, have revolutionised the state of thought and of material condition in which men live, whether they are aware of it or not. It is to this latter category that Darwin's *Origin of Species* belongs.

In Charles Darwin
The Origin of Species
Sixth Edition
Preface

Slosson, E.E.
The Book of Nature is issued only in uncut editions, and the scientist has to open its pages one by one as he reads.

Keeping Up with Science
Introduction (p. vi)

BOTANIST

Butler, Samuel
Why should the botanist, geologist or other-ist give himself such airs... Is it because he names his plants or specimens with Latin names, and divides them into genre and species...

In Geoffrey Keynes and Brian Hill (eds)
Samuel Butler's Notebooks
Botanists and Draper's Shopman (p. 264)

Croll, Oswald
Oh that the Botanists of our time, who being ignorant of the internal Form of plants, know only their matter, substance, and body, would devote as much care to the discernment of the Signatures of Plants as they do to their manifold and frequently frivolous disputes about the accurate naming of them, it would render a much richer and more beneficial service to medicine.

Basilica Chymica
Tractatus de Signaturis (p. 1)

Crothers, Samuel McChord
Here are botanists who love the growing things in the fields and woods better than the specimens in their herbariums. They love to describe better than to analyze. Now and then one may meet a renegade who carries a geologist's hammer. It is a sheer hypocrisy, like a fishing rod in the hands of a contemplative rambler. It is merely an excuse for being out of doors and among the mountains.

The Gentle Reader
The Hinter-Land of Science (pp. 236–7)

Teale, Edwin Way
Today I had lunch in the city with two scientists, a botanist and an ichthyologist. The botanist said he never kept a garden and the ichthyologist said he never went fishing.

Circle of the Seasons
December 8 (p. 282)

Unknown
We botanists cannot be so mathematically exact as geographers, and where an isthmus is very narrow, we must class the peninsula with the island. How often does it happen that two large orders, say of five hundred to two thousand or three thousand species, totally distinct from each other in all these species by a series of constant characters, are yet connected by some small isolated genus of a dozen, half a dozen, nay a single species in which these characters are so inconstant, uncertain or variously combined as to leave no room for the strait, through which we ought to navigate between the two islands.

London Journal of Botany
De Candolle's Prodromus (p. 232)
Volume IX, 1845

von Linne, Carl
To you, my dearly-beloved botanists, I submit my rules, the rules which I have laid down for myself, and in accordance with which I intend to walk. If they seem to you worthy, let them be used by you also; if not, please propound something better!

Critica Botanica
Preface (pp. xxiii-xxiv)

BOTANY

Bierce, Ambrose
BOTANY, *n*. The science of vegetables—those that are not good to eat, as well as those that are. It deals largely with their flowers, which are commonly badly designed, inartistic in color, and ill-smelling.

The Devil's Dictionary

Burroughs, John
We study botany so hard that we miss the charm of the flower entirely.

The Atlantic Monthly
In the Noon of Science (p. 324)
Volume CX, September 1912

Corner, E.H.J.
Botany needs help from the tropics. Its big plants will engender big thinking.

In Margaret D. Lowman
Life in the Treetops
Introduction (p. 1)

Dickens, Charles
When he has learnt that bottinney means a knowledge of plants, he goes and knows 'em. That's our system, Nickleby; what do you think of it?

Nicholas Nickleby
Chapter VIII (p. 114)

Dickinson, Emily
I pull a flower from the woods,—
A monster with a glass
Computes the Stamens in a breath,
And has her in a class.

Poems
Second Series
Old-Fashioned

It is foolish to call them 'flowers,'
Need the wiser tell?
If the savans 'classify' them,
It is just as well!

Poems (1890–1896)
Third series
XI

Einstein, Albert
One ought to be ashamed to make use of the wonders of science embodied
in a radio set, the while appreciating them as little as a cow appreciates the
botanic marvels in the plants she munches.

Cosmic Religion
On Radio (p. 93)

Emerson, Ralph Waldo
Love not the flower they pluck, and know it not,
And all their botany is Latin names.

Collected Poems and Translations
Blight

Esar, Evan
[Botany] The only thing about flowers that coeds dislike.

Esar's Comic Dictionary

Henslow, John Stevens
To obtain a knowledge of a science of observation, like botany, we need
make very little more exertion at first than is required for adapting
a chosen set of terms to certain appearances of which the eye takes
cognisance, and when this has been attained, all the rest is very much like
reading a book after we have learned to spell, where every page affords a
fresh field of intellectual enjoyment.

Magazine of Zoology and Botany
On the Requisites Necessary for the Advance of Botany (p. 115)
Volume 1, 1837

James, Henry
After much labor bestowed on botany, and many volumes composed
on that subject, it appears very little advanced above infancy: no other
science has made so slow a progress...[The study of botany] has been
mostly confined to giving names to plants, and to distribute them into
classes; not by distinguishing their powers and properties, but by certain

visible marks. This is an excellent preparation for composing a dictionary: but it leaves us in the dark as to the higher parts of the science...

The Gentleman Farmer
Appendix
Article IV (p. 393)

Jefferson, Thomas

And botany I rank with the most valuable sciences, whether we consider its subjects as furnishing the principal subsistence of life to man and beast, delicious varieties for our tables, refreshment from our orchards, the adornments of our flower-borders, shade and perfume of our groves, materials for our buildings or medicaments for our bodies...

Nature Magazine
In Eva Beard
Thomas Jefferson, Statesman and Scientist (p. 202)
April 1958

Mavor, William

There are few studies more cultivated at present by persons of taste, than Botany; and certainly, of all those not immediately conducive to the wants of society and the necessities of life, none can be more deserving of regard. Whether we consider the effects of Botany as enlarging the sphere of knowledge, or as conducive to health and innocent amusement, it ought to rank very high in the scale of elegant acquirements.

The Lady's and Gentleman's Botanical Pocket Book

Murray, Charlotte

The expensive apparatus of the Observatory, and the Labours of Chemistry, confine the science of Astronomy, and the study of Minerals to a few; whilst the research into the animal kingdom is attended with many obstacles which prevent its general adoption, and preclude minute investigation; but the study of Botany, that science by means of which we discriminate and distinguish one plant from another, is open to almost every curious mind; the Garden and the Field offer a constant source of unwearying amusement, easily obtained, and conducing to health, by affording a continual and engaging motive for air and exercise.

The British Garden

Queneau, Raymond

After nearly taking root under a heliotrope, I managed to graft myself on to a vernal speedwell where my hips and haws were squashed indiscriminately and where there was an overpowering axillary scent. There I ran to earth a young blade or garden pansy whose stalk had run to seed and whose nut, cabbage or pumpkin was surmounted by a capsule encircled by snakeweed. This corny, creeping sucker, transpiring at the palms, nettled a common elder who started to tread his daisies and give him the edge of his bristly ox-tongue, so the sensitive plant stalked off and parked himself. Two hours later, in fresh woods and pastures new, I saw this specimen again with another willowy young parasite who was shooting a line, recommending the sap to switch the top bulbous vegetable ivory element of his mantle blue to a more elevated apex—as an exercise in style.

Exercises in Style
Botanical (p. 171)

Thomson, William

Forty years ago I asked Liebig walking somewhere in the country, if he believed that the grass and flowers which we saw around us grew by mere chemical forces; he answered, "NO, no more than I could believe that a book of botany describing them grew by mere chemical force."

In P. Thompson
The Life of William Thomson
Volume II
Letter to The Times
May 2, 1903 (pp. 1099–100)

von Linne, Carl

What toils, what science would be more wearisome and painful than Botany, did not some singular spell of desire, which I myself cannot define, often hurry us into this pursuit, so that the love of plants often overcomes our self-love? Good God! When I observe the fate of Botanists, upon my word I doubt whether to call them sane or mad in their devotion to plants.

Critica Botanica
Generic Names (p. 65)

Wordsworth, William

Physician art thou? one, all eyes,
Philosopher! a fingering slave,
One that would peep and botanize
Upon his mother's grave?

The Complete Poetical Works of William Wordsworth
A Poet's Epitaph
Stanza V

BUG

Glover, Townend
From red-bugs and bed-bugs, from sand-flies and land-flies,
Mosquitoes, gallnippers and fleas,
From hog-ticks and dog-ticks, from hen-lice and men-lice,
We pray thee, good Lord, give us ease.

In Arnold Mallis
American Entomologist
Chapter 3 (pp. 64–5)

Holland, W.J.
When the moon shall have faded out from the sky, and the sun shall shine at noonday a dull cherry-red, and the seas shall be frozen over, and the ice-cap shall have crept downward to the equator from either pole, and no keels shall cut the waters, nor wheels turn in mills, when all cities shall have long been dead and crumbled into dust, and all life shall be on the very last verge of extinction on this globe; then, on a bit of lichen, growing on the bald rocks beside the eternal snows of Panama, shall be seated a tiny insect, preening its antenna in the glow of the worn-out sun, representing the sole survival of animal life on this our earth,—a melancholy "bug."

The Moth Book
The End (p. 445)

Lee, Marion
I never wanted to be a bug
Until I found one safe and snug
In the velvet heart of a pale pink rose
With petals tucked about his toes.

Lap of Luxury
Source unknown

Prelutsky, Jack
Bugs! Bugs!
I love bugs,
Yes I truly do,
Great big pink ones,
Little green stink ones,
Yellow bugs and blue.
I put you in my pockets,
And I wear you in my hair.
You are my close companions,
I take you everywhere.

A Pizza the Size of the Sun
Bugs! Bugs!

CELL

Bastin, Ted

As far as one can judge at all, the cell cannot be understood in its behavior as the basis of events at the molecular level. One would judge this because the control processes of detailed cell physiology seem to proliferate endlessly in the sense that the more one understands a given chain of reactions and their associated background dynamics, the larger is the number of ancillary, trigger and other processes which it seems necessary to call in to achieve completeness of explanation and a self-contained causal scheme.

In A.R. Peacock
Zygon
Reductionism: A Review of the Epistemological Issues and
Their Relevance to Biology and the Problem of Consciousness (p. 327)
Volume 11, Number 4, 4 December 1976

Bateson, William

When I look at a dividing cell I feel as an astronomer might do if he beheld the formation of a double star: that an original act of creation is taking place before me.

In Ruth Moore
The Coil of Life
Chapter VIII (p. 162)

Benchley, Robert

The scene is a plateau of primeval ooze. Things are in terrible shape. Nobody knows what to do because there is nobody. The Earth is practically new and nothing is alive except a lot of—what shall we say?

Two of these emerge from the mud together and sit down on a dry spot. There seems to be some sort of talking things over.

20,000 Leagues Under the Sea or David Copperfield
It Seems There Were a Couple of Cells (p. 176)

Cudmore, L.L. Larison

Our cells, the ones we love, are repositories of such fantastic architectural flights—pleasure domes far beyond even the most opiated dreams of Coleridge, a Xanadu percolating with the directed chaos of those hundreds of thousands of simultaneous chemical reactions that are life.

The Center of Life
The Universal Cell (p. 5)

Some cells are extremely visible—the egg of an ostrich, of a hen or puffin. But we cell biologists see these the way anyone would, as a large globe of yellow yolk surrounded by a transparent glutinous mass; interesting only by virtue of their behavior in soufflé or omelet.

The Center of Life
The Universal Cell (p. 5)

Cells let us walk, talk, think, make love and realize the bath water is cold.

The Center of Life
The Universal Cell (p. 6)

A cell always leaves the same first impression. It is incredibly crowded in there; a welter of structures crammed together like rush-hour riders in Tokyo or New York subways, with no apparent breathing space.

The Center of Life
Cellular Evolution (p. 50)

Delbrück, Max

The closer one looks at these performances of matter in living organisms, the more impressive the show becomes. The meanest living cell becomes a magic puzzle box full of elaborate and changing molecules, and far outstrips all chemical laboratories of man in the skill of organic synthesis performed with expedition and good judgment of balance... [A]ny living cell carries with it the experience of a billion years of experimentation by its ancestors. You cannot expect to explain so wise an old bird in a few simple words.

Transactions of the Connecticut Academy of Sciences
A Physicist Looks at Biology (p. 191)
Volume 38, 1949

Reichenbach, Hans

The production of just one living cell from inorganic matter is the most urgent problem which concerns the biologist who wants to make the theory of evolution complete... Presumably, biologists will someday construct synthetic albumen molecules of the gene type and of the protoplasm type, put them together, and thus produce an aggregate which possesses all the characteristics of a living cell. Should the experiment succeed, it would demonstrate conclusively that the origin of life can be traced back to inorganic matter.

The Rise of Scientific Philosophy
Chapter 12 (p. 202)

Rubin, Harry

... we cannot disrupt the cell to understand its living behavior because in doing so we destroy the very property we wish to understand...

Cancer Research
Cancer as a Developmental Disorder (p. 2940)
Volume 45, July 1985

Sherrington, Sir Charles

Essential for any conception of the cell is that it is no static system. It is dynamic. It is energy-cycles, suites of oxidation and reduction, concatenated ferment-actions. It is like a magic hive the walls of whose chambered spongework are shifting veils of ordered molecules, and rend and renew as operations rise and cease. A world of surfaces and streams. We seem to watch battalions of specific catalysts, like Maxwell's "demons," lined up, each waiting, stop-watch in hand, for its moment to play the part assigned to it. Yet each step is understandable chemistry.

Man on his Nature
Chapter III (p. 80)

Szent-Györgyi, Albert

The cell knows but one fuel:—hydrogen.

In Kenneth Thimann
The Life of Bacteria
Chapter V (p. 167)

Thomas, Lewis

The uniformity of earth's life, more astonishing than its diversity, is accountable by the high probability that we derived, originally, from some single cell, fertilized in a bolt of lightning as the earth cooled. It is from the progeny of this parent cell that we all take our looks; we still share genes around, and the resemblance of the enzymes of grasses to those of whales is in fact a family resemblance.

The Lives of a Cell: Notes of a Biology Watcher
The Lives of a Cell (p. 5)

CHANCE

Crick, Francis Harry Compton
When times get tough, true novelty is needed—novelty whose important features cannot be preplanned—and for this we must rely on chance. *Chance is the only source of true novelty.*

Life Itself
Chapter 4 (p. 58)

Darwin, Charles
When we look at the plants and bushes clothing an entangled bank, we are tempted to attribute their proportional numbers and kinds to what we call chance. But how false a view is this!

The Origin of Species
Chapter III
Complex Relations of all Animals and Plants to
Each Other in the Struggle for Existence (p. 37)

I am inclined to look at everything as resulting from designed laws, with the details, whether good or bad, left to the working out of what we may call chance.

In Francis Darwin (ed.)
The Life and Letters of Charles Darwin
Volume II
Darwin to Gray
22 May, 1860 (p. 104)

I cannot think that the world as we see it is the result of chance; and yet I cannot look at each separate thing as the result of Design...I am, and shall ever remain, in a hopeless muddle.

In Francis Darwin (ed.)
The Life and Letters of Charles Darwin
Volume II
Darwin to Gray
26 November, 1860

du Nouy, Lecomte

The laws of chance have rendered, and will continue to render, immense services to science. It is inconceivable that we could do without them, but they only express an admirable, subjective interpretation of certain inorganic phenomena and of their evolution. They are not a true explanation of objective reality.

Human Destiny
Chapter 3 (p. 37)

Keosian, J.

The materialist theory of the origin of life from inanimate beginnings recognizes the role of chance in the interactions of matter in the universe, but views the overall developments as in no way accidental; on the contrary, it is looked upon as an inevitable, almost inexorable, outcome of the emergence and operation of natural laws.

In D.L. Rohlfing and A.I. Oparin (eds.)
Molecular Evolution: Prebiological and Biological
The Origin of Life Problem—A Brief Critique (p. 14)

LaPlace, Pierre Simon

...chance has not reality in itself; it is only a term fit to designate our ignorance concerning the manner in which the different parts of a phenomenon are arranged among themselves and in relation to the rest of Nature.

In K.M. Baker
Condorcet: From Natural Philosophy to Social Mathematics
Chapter 3 (p. 168)

Monod, Jacques

...chance *alone* is at the source of every innovation, of all creation in the biosphere. Pure chance, absolutely free but blind, at the very root of the stupendous edifice of evolution: this central concept of modern biology is no longer one among other possible or even conceivable hypotheses. It is today the *sole* conceivable hypothesis, the only one that squares with observed and tested fact...There is no scientific concept, in any of the sciences, more destructive of anthropocentrism than this one, and no other so arouses an instinctive protest from the intensely teleonomic creatures that we are.

Chance and Necessity
Chapter VI (pp. 112–13)

Reichenbach, Hans
Like pebbles on the beach, biological species are ordered through a selective cause; chance in combination with selection produces order.

The Rise of Scientific Philosophy
Chapter 12 (p. 199)

Thoreau, Henry David
How many things are now at loose ends! Who knows which way the wind will blow tomorrow.

The Writings of Henry David Thoreau
Volume IV
Paradise (To Be) Regained (p. 283)

CHAOS

Adams, Henry Brooks
...Chaos was the law of nature; Order was the dream of man.

The Education of Henry Adams
The Grammar of Science (p. 420)

Briefly chaos is all that science can logically assert of the supersensuous.

The Education of Henry Adams
The Grammar of Science (p. 420)

Blackie, John Stuart
Chaos, Chaos, infinite wonder!
Wheeling and reeling on wavering wings;...

Musa Burschicosa
A Song of Geology
Second stanza

Kant, Immanuel
...God has put a secret art into the forces of nature so as to enable it to fashion itself out of chaos into a perfect world system...

Universal Natural History and Theory of the Heavens
Preface (p. 27)

Santayana, George
Chaos is perhaps at the bottom of everything: which would explain why perfect order is so rare and precarious.

Dominations and Powers
First Book, Part 1, Chapter 1 (p. 33)

Chaos is a name for any order that produces confusion in our minds.

Dominations and Powers
First Book, Part 1, Chapter 1 (p. 33)

Wilde, Oscar
Is this the end of all that primal force
Which, in its changes being still the same,
From eyeless Chaos cleft its upward course,
Through ravenous seas and whirling rocks and flame,
Till the suns met in heaven and began
Their cycles, and the morning stars sang, and the Word was Man!

Poems
Humanitad, Stanza 72

CHARACTERISTICS

Ardrey, Robert
...acquired characteristics cannot be inherited, and that within a species every member is born in the essential image of the first of its kind.

African Genesis
Chapter I, Section 2 (p. 12)

CHROMOSOMES

Conklin, E.G.
What molecules and atoms and electrons are to the physicist and chemist, chromosomes and genes are to the biologist.

Science
A Generation's Progress in the Study of Evolution (p. 151)
Volume 80, Number 2068, August 17, 1934

Newman, Joseph S.
All living protoplasmic cells
That make up frogs or pimpernels
Or men or hippopotami
Have portions known as nuclei.
Within these microscopic homes
There lurk our fateful chromosomes,
Those strange hereditary factors
That make us good or bad actors,
That shape our lips and chins and eyebrows
And predetermine fools and highbrows.

Poems for Penguins
Heredity

Schrödinger, Erwin
The chromosome structures are at the same time instrumental in bringing about the development they foreshadow. They are law-code and executive power—or, to use another simile, they are architect's plan and builder's craft—in one.

What Is Life?
Chapter II, Section 12 (p. 21)

Stoller, Robert
What to the unempathic scientist is a chromosome is the heavy hand of immutable destiny to the victims: on receiving the genetic information,

103

the patient may feel transformed into a freak, no longer fully human. Those who feel this is an exaggeration have not treated people afflicted with depression, hopelessness, or psychosis as a result of learning such a truth.

In Michael A. Sperber and Lissy F. Jarvik
Psychiatry and Genetics
Genetics, Constitution, and Gender Disorder (p. 54)

CLASSIFICATION

Agassiz, Louis
Are these divisions artificial or natural? Are they the devices of the human mind to classify and arrange our knowledge in such a manner as to bring it more readily within our grasp and facilitate further investigations, or have they been instituted by the Divine Intelligence as the categories of his mode of thinking?

Essay on Classification
Chapter I, Section I (p. 8)

Bronowski, Jacob
It is not obviously silly to classify flowers by their colors; after all, the bluer flowers do tend to be associated with the colder climates and greater heights. There is nothing wrong with the system in advance. It simply does not work as conveniently and as instructively as Linnaeus's classification by family likenesses.

The Common Sense of Science
Chapter IV, Section 4 (p. 48)

Emerson, Ralph Waldo
But what is classification but the perceiving that these objects are not chaotic, and are not foreign, but have a law which is also the law of the human mind?

The Collected Works of Ralph Waldo Emerson
Volume I
The American Scholar (p. 54)

Genesis 2:20
And Adam gave names to all cattle, and to the fowl of the air, and to every beast.

The Bible

Graton, L.C.
The purpose of classification is not to set forth final and indisputable truths but rather to afford stepping stones towards better understanding.

In Fred M. Bullard
Volcanoes of the Earth
Chapter 4 (p. 30)

Hopwood, A. Tindell
The urge to classify is a fundamental human instinct; like the predisposition to sin, it accompanies us into the world at birth and stays with us to the end.

Proceedings of the Linnean Society of London
The Development of Pre-Linnaean Taxonomy (p. 230)
Volume 170, 1959

James, William
The first steps in most of the sciences are purely classificatory. Where facts fall easily into rich and intricate series (as plants and animals and chemical compounds do), the mere sight of the series fill the mind with a satisfaction *sui generis*; and a world whose *real* materials naturally lend themselves to serial classification is *pro tanto* a more rational world, a world with which the mind will feel more intimate, than with a world in which they do not. By the pre-evolutionary naturalists, whose generation has hardly passed away, classifications were supposed to be ultimate insights into God's mind, filling us with adoration of his ways. The fact that Nature lets us make them was a proof of the presence of his Thought in her bosom.

The Principles of Psychology
Volume II
Necessary Truths—Effects of Experience
Classificatory Series (p. 647)

Morris, H.M.
If an evolutionary continuum existed, as the evolution model should predict, there would be no gaps, and thus it would be impossible to demark specific categories of life. Classification requires not only similarities, but differences and gaps as well, and these are much more amenable to the creation model.

Scientific Creationism
Chapter IV (p. 72)

Olson, S.L.
...the present classification of birds amounts to little more than superstition and bears about as much relationship to a true phylogeny of the Class *Aves* as Greek mythology does to the theory of relativity.

The Auk
The Museum Tradition in Ornithology. A Response to Ricklefs (p. 193)
Volume 98, January 1981

Pope, Alexander
Where order in variety we see
And where, though all things differ, all agree.

Alexander Pope's Complete Poems
Windsor Forest, l. 15–16

Smiles, Samuel
A place for everything and everything in its place.

Thrift
Chapter 5 (p. 66)

CLONE

Ehlers, Vernon

Human life is sacred. The good Lord ordained a time-honored method of creating human life, commensurate with substantial responsibility on the part of the parents, the responsibility to raise a child appropriately. Creating life in the laboratory is totally inappropriate and so far removed from the process of marriage and parenting that has been instituted upon this planet that we must rebel against the very concept of human cloning. It is simply wrong to experiment with the creation of human life in this way.

Congressional Record-House
Human Cloning (p. H 713)
Volume 143, No. 26, 4 March 1997

Unknown

Mary had a little lamb, its fleece was slightly gray
It didn't have a father, just some borrowed DNA.

It sort of had a mother, though the ovum was on loan,
It was not so much a lambkin as a little lamby clone.

And soon it had a fellow clone, and soon it had some more,
They followed her to school one day, all cramming through the door.

It made the children laugh and sing, the teachers found it droll,
There were too many lamby clones, for Mary to control.

No other could control the sheep, since the programs didn't vary
So the scientists resolved it all, by simply cloning Mary.

But now they feel quite sheepish, those scientists unwary,
One problem solved but what to do, with Mary, Mary, Mary.

Posted On the Internet

COLEOPTERIST

Crowson, Roy A.

If and when the day comes when pure science is once again generally appreciated as a self-justifying intellectual adventure of mankind, then the coleopterists should be able to step forward and claim their share of its glory.

The Biology of the Coleoptera
Chapter 21 (p. 691)

COLLECTING

Durrell, Gerald M.
One of the chief charms of collecting is its uncertainty. One day you will go out loaded down with nets and bags for the sole purpose of catching bats, and you will arrive back in camp with a python in the nets, your bags full of birds, and your pockets full of giant millipedes.

The Overloaded Ark
Chapter 5 (p. 92)

CONCHOLOGY

Garstang, Walter

Echinospira sets this riddle to the students of Conchology
To make them pay attention to the doctrines of Morphology:
And this is how he poses it: "The Ammonite's old shell
From time to time was portioned off, to make it fit him well.

"The smaller shell around his hump was 'visceral,' like mine;
His outer shell also agrees: it's 'pallial' in fine.
We differ in this: his inner shell was fixed by suture,
While mine is truly portable, and useful for the future!

So let us sing in fitting terms an *entente cordiale*,
Observing in its proper place the *torsion viscerale*:
My outer shell's a 'relic' of my Ammonitic traits,
My inner is a tribute to my clever modern ways!"

Larval Forms
Echinospira's Double Shell
Stanza 2–4 (p. 42)

CONSERVATION

Carson, Rachel
Our attitudes toward plants is a singularly narrow one. If we see any immediate utility in a plant we foster it. If for any reason we find its presence undesirable or merely a matter of indifference, we may condemn it to destruction forthwith.

Silent Spring
Chapter 6 (p. 63)

Leopold, Aldo
To keep every cog and wheel is the first precaution of intelligent tinkering.
A Sand County Almanac: With Essays on Conservation From Round River
Round River (p. 190)

Conservation is a state of harmony between men and land.

A Sand County Almanac: With Essays on Conservation From Round River
The Land Ethic (p. 207)

Lovejoy, Thomas E.
In the last analysis, even when we have learned to manage other aspects of the global environment, even if population reaches a stable level, even if we reach a time when environmental crises have become history, even if most wastes have gone except the most long lived, even if global cycles have settled back into more normal modes, then the best measurement of how we have managed the global environment will be how much biological diversity has survived.

In D.B. Botkin, M.F. Caswell, J.E. Estes and A.A. Orio (eds)
Changing the Global Environment: Perspectives on Human Involvement
Deforestation and Extinction of Species (p. 97)

Osborn, Henry Fairfield
[The] great battle for preservation and conservation cannot be won by gentle tones, nor by appeals to the aesthetic instincts of those who have no sense of beauty, or enjoyment of Nature.

In William T. Hornaday
Our Vanishing Wild Life
Preface (p. vii)

Sheldrick, Daphne
With amazing arrogance we presume omniscience and an understanding of the complexities of Nature, and with amazing impertinence we firmly believe that we can better it... [W]e have forgotten that we, ourselves, are just a part of nature, an animal which seems to have taken the wrong turning bent on total destruction.

The Tsavo Story
Chapter 15 (p. 190)

CONSULTANTS

Grindal, Bruce
Salamone, Frank
In the past twenty years, "doing" anthropology has become more and more complex. In the days when we traveled long distances to far-off places, our fieldwork stayed in the field. Now, the distances have been narrowed. Informants have become consultants. Consultants are our friends. As such, they can board a plane in their land and come to visit, spending long nights in earnest conversation about truth and meaning and enlightenment and expectations. In the days when we wrote only inscrutable manuscripts circulated among colleagues, there was no one to dispute the validity of our work except another "expert" in the area. Now, our consultant-friends are critics, editors of our written words, commentators of their lives, and ours.

Bridges to Humanity (p. 193)

CREATIONISM

Cloud, Preston

Fundamentalist creationism is not a science but a form of antiscience, where more vocal practitioners, despite their master's and doctoral degrees in the sciences, play fast and loose with the facts of geology and biology.

In J. Peter Zetterberg (ed.)
Evolution versus Creationism (p. 134)

Gould, Stephen Jay

The argument that the literal story of Genesis can qualify as science collapses on three major grounds: the creationists' need to invoke miracles in order to compress the events of the earth's history into the biblical span of a few thousand years; their unwillingness to abandon claims clearly disproved, including the assertion that all fossils are products of Noah's flood; and their reliance upon distortion, misquote, half-quote, and citation out of context to characterize the ideas of their opponents.

The Skeptical Inquirer
The Verdict on Creationism (p. 186)
Volume 12, Winter 87/88

"Creation science" has not entered the curriculum for a reason so simple and so basic that we often forget to mention it: because it is false, and because good teachers understand exactly why it is false. What could be more destructive of that most fragile yet most precious commodity in our entire intellectual heritage—good teaching—than a bill forcing honorable teachers to sully their sacred trust by granting equal treatment to a doctrine not only known to be false, but calculated to undermine any general understanding of science as an enterprise?

The Skeptical Inquirer
The Verdict on Creationism (p. 186)
Volume 12, Winter 87/88

Laudan, Larry
Rather than taking on the creationists obliquely and in wholesale fashion by suggesting that what they are doing is "unscientific" *tout court* (which is doubly silly because few authors can even agree on what makes an activity scientific), we should confront their claims directly and in piecemeal fashion by asking what evidence and arguments can be marshaled for and against each of them. The core issue is not whether Creationism satisfies some undemanding and highly controversial definition of what is scientific; the real question is whether the existing evidence provides stronger arguments for evolutionary theory than for Creationism.

Science, Technology & Human Values
Commentary: Science at the Bar—Cause for Concern (p. 18)
Volume 7, Number 41, Fall 1982

Lyell, Charles
Whatever be the power which has for hundreds of times repeopled the Earth with tribes of plants & animals as fast as they became extinct, that power I have always held is still in full & unabated action as is its antagonist or destructive power.

In Leonard G. Wilson (ed.)
Sir Charles Lyell's Scientific Journals on the Species Question
Journal II
July 10, 1856 (p. 124)

Moore, John A.
It becomes evermore important to understand what is science and what is not. Somehow we have failed to let our students in on that secret. We find as a consequence, that we have a large and effective group of creationists who seek to scuttle the basic concept of the science of biology—the science that is essential for medicine, agriculture, and life itself; a huge majority of citizens who, in "fairness," opt for presenting as equals the "science" of creation and the science of evolutionary biology; and a president who is so poorly informed that he believes that scientists are questioning that evolution ever occurred. It is hard to think of a more terrible indictment of the way we have taught science.

In J. Peter Zetterberg (ed.)
Evolution versus Creationism (p. 3)

Morris, H.M.
Creationism is consistent with the innate thoughts and daily experiences of the child and thus is conducive to his mental health. He knows, as part of his own experience of reality, that a house implies a builder and a watch a watchmaker. As he studies the still more intricately complex nature of,

say, the human body, or the ecology of a forest, it is highly unnatural for him to be told to think of these systems as chance products of irrational processes.

Scientific Creationism
Chapter I (p. 14)

It seems beyond all question that such complex systems as the DNA molecule could never arise by chance, no matter how big the universe or how long the time. The creation model faces this fact realistically and postulates a great Creator, by whom came life.

Scientific Creationism
Chapter IV (p. 62)

Nelkin, Dorothy

Creationism is a 'gross perversion of scientific theory'. Scientific theory is derived from a vast mass of data and hypotheses, consistently analysed; creation theory is 'God given and unquestioned', based on an *a priori* commitment to a six-day creation. Creationists ignore the interplay between fact and theory, eagerly searching for facts to buttress their beliefs. Creationism cannot be submitted to independent testing and has no predictive value, for it is a belief system that must be accepted on faith.

Science Textbook Controversies and the Politics of Equal Time
Chapter 6 (p. 89)

Patterson, John W.

There are many facets to "scientific creationism" and the movement can be discussed in any of several ways. However, it is best viewed as a loosely connected group of fundamentalist ministries led largely by scientifically incompetent engineers.

In J. Peter Zetterberg (ed.)
Evolution versus Creationism (p. 151)

CRUSTACEAN

Pallister, William
With eight thousand CRUSTACEAN species, we list
All the lobsters and crabs, many others beside;
On the beaches and tide-strips their races subsist
On the wreck of the sea and the wrack of the tide;
In his jointed shell, hungry and seeking each goes,
If one loses a claw, soon another one grows.

Poems of Science
Beginnings

CRAB

James, William
Probably a crab would be filled with a sense of personal outrage if it could hear us class it without ado or apology as a crustacean, and thus dispose of it. "I'm no such thing," it would say, "I am MYSELF, MYSELF alone."

The Varieties of Religious Experience
Lecture I (p. 17)

CRAWFISH

Flaubert, Gustave
Crayfish.   Female of the lobster.   Walks backward.   Always call reactionaries "crayfish."

Dictionary of Accepted Ideas
Animal Life (pp. 139–40)

WOODLOUSE

Garstang, Walter

MacBride was in his garden settling pedigrees,
When came a baby Woodlouse and climbed upon his knees,
And said: "Sir, if our six legs have such an ancient air,
Shall we be less ancestral when we've grown our mother's pair?"

Larval Forms
Isopod Phylogeny
Stanza 3 (p. 50)

DARWINISM

Huxley, Julian
Darwinism removed the whole idea of God as the creator of organisms from the sphere of rational discussion.

In S. Tax and C. Callender
Issues in Evolution
Volume III
Evolution After Darwin
At Random (p. 45)

Jones, F. Wood
Only a fool could deny the revolutionary impact of Darwinism on the outlook of the nineteenth century, when—as one biologist put it—the educated public was faced with the alternative 'for Darwin or against evolution'. But the narrow sectarianism of the neo-Darwinists of our own age is an altogether different matter; and in the not-too-distant future biologists may well wonder what kind of benightedness it was that held their elders in thrall.

In Arthur Koestler
Janus: A Summing Up
Chapter X, Section 5 (p. 204)

Kitcher, Philip
Darwin is the Newton of Biology.

Abusing Science
Chapter 2 (p. 54)

McKibben, Bill
"Science," of course, replaced "God" as a guiding concept for many people after Darwin. Or, really, the two were rolled up into a sticky ball. To some degree this was mindless worship of a miracle future, the pursuit of which has landed us in the fix we now inhabit.

The End of Nature
The End of Nature (pp. 80–1)

Newman, Joseph S.
What countless procreative mates
Brought plasmic cells to vertebrates
And blazed the long ancestral trails
That substituted brains for tails!
For when the human kind began
It did not spring full-blown to man;
It started from the very seed
That branched to snail and centipede,
And which, by devious ways Darwinian,
Made oyster, lobster, and Virginian.

Poems for Penguins
Anthropology

Shaw, George Bernard
…as compared with the open-eyed intelligent wanting and trying of Lamarck, the Darwinian process may be described as a chapter of accidents. As such, it seems simple, because you do not at first realise all that it involves. But when its whole significance dawns on you, your heart sinks into a heap of sand within you. There is a hideous fatalism about it, a ghastly and damnable reduction of beauty and intelligence, of strength and purpose, of honour and aspirations, to such casually picturesque changes as an avalanche may make in a mountain landscape, or a railway accident in a human figure.

Back To Methuselah
Preface
The Moment and the Man (p. xl)

Walker, Michael
One is forced to conclude that many scientists and technologists pay lip-service to Darwinian theory only because it supposedly excludes a Creator…

Quadrant
October 1982 (p. 44)

DATA

Fort, Charles
The interpretations will be mine, but the data will be for anybody to form his own opinions on.

In Damon Knight
Charles Fort: Prophet of the Unexplained
A Charles Fort Sampler (p. vii)

Jennings, H.S.
...the biologist has a more intimate access to a certain sample of his material, for he is himself that sample. Through this fact he discovers certain things about the materials of biological science that he cannot discover by the other method alone...he finds that the thing to be studied by the biologist include emotions, sensations, impulses, desires...Thus the biologist has two sets of data, discovered in somewhat different ways, one set being discoverable only through the fact that the biologist is himself a biological specimen.

The Universe and Life
Nature of the Universe (pp. 9, 10)

Morris, H.M.
The data must be *explained* by the evolutionist, but they are *predicted* by the creationist.

Scientific Creationism (p. 13)

Woodger, Joseph Henry
We are, therefore, in danger of being overwhelmed by our data and of being unable to deal with the simpler problems first and understand their connexion. The continual heaping up of data is worse than useless if interpretation does not keep pace with it. In biology this is all the more deplorable because it leads us to slur over what is characteristically biological in order to reach hypothetical 'causes.'

Biological Principles
Chapter VI (p. 318)

DEATH

Asimov, Isaac

[Death] is an essential part of the successful functioning of life...new organisms cannot perform their role properly unless the old ones are removed from the scene after they have performed their function in producing the new. In short, the death of the individual is essential to the life of the species.

A Choice of Catastrophes
Chapter 12 (p. 239)

Dawkins, Richard

...however many ways there may be of being alive, it is certain that there are vastly more ways of being dead, or rather not alive.

The Blind Watchmaker
Chapter 1 (p. 9)

Lovelock, J.E.

...the unending death-roll of all creatures, including ourselves, is the essential complement to the unceasing renewal of life.

Gaia
Chapter 8 (p. 125)

Muir, John

Leaves have their time to fall, and though indeed there is a kind of melancholy present when they, withered and dead, are plucked from their places and made the sport of the gloomy autumn wind, yet we hardly deplore their fate, because there is nothing unnatural in it. They have done all that their Creator wished them to do, and they should not remain longer in their green vigor.

In Sally M. Miller (ed.)
John Muir: Life and Work
Part I, Chapter 1 (p. 28)

Sakaki, Nanao
At a department store in Kyoto
One of my friends bought a beetle
For his son, seven years old.

A few hours later
The boy brought his dead bug
To a hardware store, asking
"Change battery please."

Break the Mirror
Future Knows (p. 27)

Strehler, Bernard
Aging and death do seem to be what Nature has planned for us. But what
if we have other plans?

In J. Lyon and P. Gorner
Altered Fates
Part II (p. 295)

Teale, Edwin Way
In nature, there is less death and destruction than death and transmutation.

Circle of the Season
July 5 (p. 143)

Thoreau, Henry David
Every part of nature teaches that the passing away of one life is the making room for another. The oak dies down to the ground, leaving within its rind a rich virgin mold, which will impart a vigorous life to an infant forest.

Journal
Volume I: 1837–44
October 24, 1837

DISCOVERY

Beveridge, W.I.B.
Probably the majority of discoveries in biology and medicine have been come upon unexpectedly, or at least had an element of chance in them, especially the most important and revolutionary ones.

The Art of Scientific Investigation
Chapter 3 (p. 31)

Hitching, Francis
Science is a voyage of discovery, and beyond each horizon there is another.

The Neck of the Giraffe
Part 3
Chapter 9 (p. 263)

Kuhn, Thomas
Discovery commences with the awareness of anomaly, i.e., with the recognition that nature has somehow violated the paradigm-induced expectations that govern normal science. It then continues with a more or less extended exploration of the area of anomaly. And it closes only when the paradigm theory has been adjusted so that the anomalous has become the expected...Until he has learned to see nature in a different way—the new fact is not quite a scientific fact at all.

The Structure of Scientific Revolutions
Chapter VI (pp. 52–3)

Lamarck, Jean Baptiste Pierre Antoine
...the most important discoveries of the laws, methods and progress of nature have nearly always sprung from the examination of the smallest objects which she contains...

Zoological Philosophy
Preliminary Discourse (pp. 9–10)

127

Shaw, George Bernard

... any fool can make a discovery. Every baby has to discover more in the first years of its life than Roger Bacon ever discovered in his laboratory.

Back to Methuselah
Tragedy of an Elderly Gentleman
Part IV
Act I
Tragedy of an Elderly Gentleman (p. 160)

Thomson, Joseph John

As we conquer peak after peak we see in front of us regions full of interest and beauty, but we do not see our goal, we do not see the horizon: in the distance tower still higher peaks, which will yield to those who ascend them still wider prospects.

In Bernard Jaffe
Crucibles
Epilogue (p. 351)

Thoreau, Henry David

Do not engage to find things as you think they are.

The Writings of Henry David Thoreau
Volume VI
Letter to Harrison Blake
August 9, 1850 (p. 186)

Twain, Mark

What is there that confers the noblest delight? What is that which swells a man's breast with pride above that which any other experience can bring to him? Discovery! To know that you are walking where none others have walked; that you are beholding what human eye has not seen before; that you are breathing a virgin atmosphere. To give birth to an idea, to discover a great thought—an intellectual nugget, right under the dust of a field that many a brain-plough had gone over before. To find a new planet, to invent a new hinge, to find a way to make the lightnings carry your message. To be the *first*—that is the idea.

The Innocents Abroad
Chapter 26 (p. 209)

Whitehead, Alfred North

The true method of discovery is like the flight of an aeroplane. It starts from the ground of particular observation; it makes a flight in the thin air of imaginative generalization; and it again lands for renewed observation rendered acute by rational interpretation.

Process and Reality
Chapter I, Section II (p. 7)

DISPERSAL

Zimmerman, E.C.
We must recognize that it is abnormal conditions that account for much overseas dispersal. It is not the soft, gentle trade wind—it is the irresistible hurricane that is the key.

In J. Linsley Gressitt (ed.)
Pacific Basin Biogeography
Pacific Basin Biogeography: A Summary Discussion (p. 478)

So many continents and land bridges have been built in and across the Pacific by biologists that, were they all plotted on a map, there would be little space left for water. Whenever a particularly puzzling problem arises, the simplest thing seems to be to build a continent or bridge, rather than to admit defeat at the hands of nature, or to consider the data at hand inadequate for solving the problem. Most of the land bridges suggested to account for the distribution of certain plants and animals in the Pacific create more problems than they solve. If the central and eastern Pacific ever included large land areas and land bridges, there should be some indication of the consequent peculiar development of the fauna and floras, but there is no such evidence.

American Naturalist
Distribution and Origin of Some Eastern Oceanic Insects (p. 282)
Volume LXXVI, Number 764, 1942

DISSECTION

Barbellion, W.N.P.

Dissected the Sea Urchin (*Echinus esculentus*). Very excited over my first view of Aristotle's Lantern. These complicated pieces of animal mechanism never smell of musty age—after aeons of evolution. When I open a Sea Urchin and see the Lantern, or dissect a Lamprey and cast eyes on the branchial basket, such structures strike me as being as finished and exquisite as if they had just a moment before been tossed me fresh from the hands of the Creator. They are fresh, young, they smell *new*.

The Journal of a Disappointed Man
November 3, 1908 (p. 19)

Butler, Samuel

As if a man should be dissected,
To see what part is disaffected.

Hudibras
Part II, Canto I (l. 505–6)

Pope, Alexander

Life following life through creatures you dissect,
You lose it in the moment you detect.

Alexander Pope's Complete Poems
Moral Essays
Epistle I, l. 29–30

Wadsworth, William

Sweet is the lore which Nature brings;
Our meddling intellect
Mis-shapes the beauteous form of things:—
We murder to dissect.

The Complete Poetical Works of William Wordsworth
The Tables Turned
The Thorn

130

DIVERGENCE

Darwin, Charles
As buds give rise by growth to fresh buds, and these, if vigorous, branch out and overtop on all sides many a feebler branch, so by generation I believe it has been with the great Tree of Life, which fills with its dead and broken branches the crust of the earth, and covers the surface with its ever branching and beautiful ramifications.

The Origin of Species
Chapter IV
Summary of Chapter (p. 64)

DNA

Baum, Harold
The primary sequence of proteins
Is coded with DNA
On sense strand of the double helix
Coiled antiparallel way.
(Introns and exons, changes post-transcriptional, and all
Glycosylations, don't alter such basics at all.)

The Biochemists' Handbook
Protein Biosynthesis
Tune: "My Bonnie Lies Over the Ocean"

Boulding, Kenneth E.
DNA was the first three-dimensional Xerox machine.

In Richard P. Beilock (ed.)
Beasts, Ballads, and Bouldingisms
Evolution, Ecology, and Spaceship Earth (p. 160)

Crick, Francis Harry Compton
Nowadays most people know what DNA is, or if they don't know it must
be a dirty word, like "chemical" or "synthetic".

What Mad Pursuit
Chapter 6 (p. 63)

Dobzhansky, Theodosius
The potentiality of mind must be present in the egg and the sperm and
in the DNA molecules. But it does not follow that eggs and sperms
themselves have minds. A stone has in it a potentiality to become a statue,
but it does not follow that every stone has a statue concealed in it.

The Biology of Ultimate Concern
Chapter 2 (p. 30)

Jukes, Thomas Hughes
Slowly the molecules enmeshed in ordered asymmetry.
A billion years passed, aeons of trial and error.
The life message took form, a spiral,
A helix, repeating itself endlessly,
Swathed in protein, nurtured by
Enzymes, sheltered in membranes,
Laved by salt water, armored with lime.

Molecules and Evolution (p. iii)

Thomas, Lewis
The greatest single achievement of nature to date was surely the invention of DNA. We have had it from the very beginning, built into the first cell to emerge, membranes and all, somewhere in the soupy waters of the cooling planet.

In N. Tiley
Discovering DNA
Introduction (p. vii)

Watson, James D.
Crick, Francis Harry Compton
We wish to suggest a structure of the salt of deoxyribosenucleic acid (D.N.A.). This structure has novel features which are of considerable biological interest.

Nature
Molecular Structure of Nucleic Acids (p. 737)
Volume 171, Number 4356, April 25, 1953

ECOLOGY

Allaby, Michael
Ecology is rather like sex—every new generation likes to think they were the first to discover it.

The Times (London)
6 October 1989

Berry, R.J.
Bradshaw, A.D.
Ecology and genetics have always been uneasy bedfellows, despite their intrinsic complementarity; genetics is about what exists, ecology is about how it exists.

In R.J. Berry, T.J. Crawford and G.M. Hewitt (eds)
Genes in Ecology
Genes in the Real World (p. 431)

Ecology lacks an agreed theoretical core and is therefore easily destabilized and subject to intellectual fashion.

In R.J. Berry, T.J. Crawford and G.M. Hewitt (eds)
Genes in Ecology
Genes in the Real World (p. 431)

Borland, Hal
The pond and the wetlands are a world unto themselves. The adventurer there, be he novice or veteran, will be aware of ancient beginnings and insistent change. There he will see those subtle interrelationships of life which the specialist calls ecology.

Beyond Your Doorstep
Chapter 5 (p. 103)

Elton, Charles
... there is more ecology in the Old Testament or the plays of Shakespeare than in most of the zoological textbooks ever printed.

Animal Ecology
Chapter II (p. 7)

At a time when ecology and genetics are each racing swiftly towards one new concept after another, yet with little contact of thought between the two subjects, there may be some advantage in surveying, if only synoptically and in preliminary fashion, the largely uncharted territory between them.

> In G.R. de Beer (ed.)
> *Evolution: Essays on Aspects of Evolutionary Biology Presented to Professor E.S. Goodrich on his Seventieth Birthday*
> Animal Numbers and Adaptation (p. 127)

Foreman, Dave

But, damn it, I am an animal. A living being of flesh and blood, storm and fury. The oceans of the Earth course through my veins, the winds of the sky fill my lungs, the very bedrock of the planet makes my bones. I am alive! I am not a machine, a mindless automaton, a cog in the industrial world, some New Age android. When a chain saw slices into the heartwood of a two-thousand-year-old Coast Redwood, it's slicing into my guts. When a bulldozer rips through the Amazon rain forest, it's ripping into my side. When a Japanese whaler fires an exploding harpoon into a great whale, my heart is blown to smithereens. I am the land, the land is me.

> *Confessions of an Eco-warrior*
> Chapter 1 (pp. 4–5)

Haeckel, Ernst

[Ecology] the science of relations between organisms and their environment.

> In Anna Bramwell
> *Ecology in the 20th Century: A History*
> Chapter 3 (p. 40)

Kühnelt, Wilhelm

The protection of an animal or of a plant will be ineffectual so long as we do not also preserve that organism's conditions of life.

> In Philippe Diolé
> *The Errant Ark*
> Chapter 3 (p. 69)

Sontag, Susan

Guns have metamorphosed into cameras in this earnest comedy, the ecology safari, because nature has ceased to be what it had always been— what people needed protection from. Now nature—tamed, endangered, mortal—needs to be protected from people. When we are afraid, we shoot. But when we are nostalgic, we take pictures.

On Photography
In Plato's Cave (p. 15)

Tansley, A.G.

Every genuine worker in science is an explorer, who is continually meeting fresh things and fresh situation, to which he has to adapt his material and mental equipment. This is conspicuously true of our subject, and is one of the greatest attractions of ecology to the student who is at once eager, imaginative, and determined. To the lover of prescribed routine methods with the certainty of 'safe' results the study of ecology is not to be recommended.

Practical Plant Ecology (p. 97)

Ward, Barbara

We cannot cheat on DNA. We cannot get round photosynthesis. We cannot say I am not going to give a damn about phytoplankton. All these tiny mechanisms provide the preconditions of our planetary life. To say we do not care is to say in the most literal sense that "we choose death."

Who Speaks for Earth?
Speech for Stockholm (p. 31)

ENTOMOLOGY

Evans, Howard Ensign
If insects were the size of birds, or people the size of mice, "bug watchers" would be as prevalent as bird watchers, and entomologists would command the budget of the Defense Department. But as it is, entomologists have a good deal of trouble explaining what their science is all about, or for that matter how it is spelled.

The Pleasures of Entomology
Preface (p. 9)

Holmes, Oliver Wendell
I suppose you are an entomologist?—I said with a note of interrogation.

—Not quite so ambitious as that, sir. I should like to put my eye on the individual entitled to that name! A *society* may call itself an Entomological Society, but a man who arrogates such a broad title as that to himself, in the present state of science, is a pretender sir, a dilettante, an imposter! No man can be truly called an entomologist, sir; the subject is too vast for any single human intelligence to grasp.

The Poet at the Breakfast-Table
Chapter II (p. 49)

Howard, Leland O.
People think entomologists have small minds because they interest themselves in small animals.

In Edwin Teale
Circle of the Seasons
February 19 (p. 34)

Kirby, William
Spence, William
...in the minds of most men...an *Entomologist* is synonymous with everything futile and childish. [Involved in a] science which, in nine

companies out of ten companies with which he may associate, promises
to signalise him as an object of pity or contempt.

An Introduction to Entomology
Preface to the First Edition (p. ix)

IS HE LOOKING AT YOU OR ME?

Nash, Ogden

He was an eminent etymologist, which is to say he knew nothing but bugs.
He could tell the Coleoptera from the Lepidoptera,
And the Aphidae and the Katydididae from the Grasshoptera.

Verses from 1929 On
The Strange Case of the Entomologist's Heart

Unknown

A gentle reader drop a tear
For one beneath this stone
In life he named 7,000 bugs
To science, all unknown.

But now, alack! He is condemned

In a place I dare not name
With his own books, through endless years
To identify the same.

Entomological News
Obituary of an Entomologist (p. 297)
Volume 13, Number 9, 1902

Wood, John George
The study of entomology is one of the most fascinating of pursuits. It takes its votaries into the treasure-house of Nature, and explains some of the wonderful series of links which form the great chain of creation. It lays open before us another world, of which we have been hitherto unconscious, and shows us that the tiniest insect, so small perhaps that the unaided eye can scarcely see it, has its work to do in the world, and does it.

Source unknown

Wood, Robert William
The Plover and the Clover can be told apart with ease,
By paying close attention to the habit of the Bees,
For En-to-molo-gists aver, the Bee can be in Clover,
While Ety-molo-gists concur, there is no B in Plover.

How to Tell the Birds from the Flowers and Other Wood-cuts
The Clover. The Plover. (p. 3)

ENTROPY

Jungck, J.R.

...entropy will not be the nemesis of evolution; on the contrary, the selection of entropy-driven processes in biological systems has been responsible for the evolution of the sophisticated organization of contemporary biota.

In D.L. Rohlfing and A.I. Oparin (eds)
Molecular Evolution: Prebiological and Biological
Thermodynamics and Self Assembly:
An Empirical Example Relating to Entropy and Evolution (p. 107)

von Neumann, John

You should call it entropy for two reasons. In the first place your uncertainty function has been used in statistical mechanics under that name, so it already has a name. In the second place, and more important, "no one knows what entropy really is, so in a debate you will always have the advantage".

In M. Tribus and E.C. McIrvine
Scientific American
Energy and Information (p. 179)
Volume 225, Number 3, 1971

ENVIRONMENT

Bartram, William
This world, as a glorious apartment of the boundless palace of the Sovereign Creator, is furnished with an infinite variety of animated scenes, inexpressibly beautiful and pleasing, equally free to the inspection and enjoyment of all his creatures.

Travels and Other Writings
Introduction (p. 13)

Commoner, Barry
The environment makes up a huge, enormously complex living machine that forms a thin dynamic layer on the earth's surface, and every human activity depends on the integrity and the proper functioning of this machine. Without the photosynthetic activity of green plants, there would be no oxygen for our engines, smelters, and furnaces, let alone support for human and animal life. Without the action of the plants, animals, and microorganisms that live in them, we could have no pure water in our lakes and rivers. Without the biological processes that have gone on in the soil for thousands of years, we could have neither food crops, oil, nor coal. This machine is our biological capital, the basic apparatus on which our total productivity depends. If we destroy it, our most advanced technology will become useless and any economic and political system that depends on it will founder. The environmental crisis is a signal of this approaching catastrophe.

The Closing Circle: Nature, Man & Technology
Chapter 2 (p. 13)

Dubos, René J.
Each cell, each living being has a multipotential biochemical personality but the physiochemical environment determines the one under which it manifests itself.

Louis Pasteur
Chapter XIII (p. 383)

Elton, Charles

It is usual to speak of an animal as living in a certain physical and chemical environment, but it should always be remembered that strictly speaking we cannot say exactly where the animal ends and the environment begins—unless it is dead, in which case it has ceased to be a proper animal at all...

Animal Ecology
Chapter IV (p. 34)

Morrison, Jim

What have they done to the earth?
What have they done to our fair sister?
Ravaged and plundered,
And ripped her and bit her,
Stuck her with knives in the side of the dawn,
And tied her with fences and dragged her down.

When the Music's Over

Rickover, Hyman G.

It is a profound mistake to think of land only in terms of its money values and, however natural it may be for individuals to do this, the nation or state should never do so. It should instead act always to preserve, foster, and cause to be developed to the maximum of its capacity not the monetary, but the real and physical value of every acre of its soil, both rural and urban. This is its educative, esthetic, and, in the fullest and widest sense of the meaning, productive, creative and enduring worth.

Testimony
House Appropriations Defense Subcommittee
June 19, 1973

Snyder, Gary

A properly radical environmentalist position is in no way antihuman. We grasp the pain of the human condition in its full complexity, and add the awareness of how desperately endangered certain key species and habitats have become... The critical argument now within environmental circles is between those who operate from a human-centered resource management mentality and those whose values reflect an awareness of the whole of nature.

The Practice of the Wild
Survival and Sacrament (p. 181)

ERROR

Darwin, Charles
To kill an error is as good a service as, and sometimes even better than, the establishing of a new truth or fact.

In Francis Darwin (ed.)
More Letters of Charles Darwin
Volume II
Darwin to Wilson (p. 422)
March 5, 1879

Fischer, Ernst Peter
The way to wisdom, I explain,
Is easy to express,
To err and err and err again
But less and less and less.

Beauty and the Beast
Chapter 5 (p. 93)

Mach, Ernst
We err when we expect more enlightenment from an hypothesis than from the facts themselves.

The Science of Mechanics
Chapter V (p. 600)

Shakespeare, William
The error of our eye directs our mind.
What error leads must err...

Troilus and Cressida
Act V, Scene II, l. 110–11

ETHICS

Caplan, Arthur
The use of fetuses as organ and tissue donors is a ticking time bomb of bioethics.

Time
In Joe Levine
Help from the Unborn (p. 62)
Volume 129, Number 2, January 12, 1987

Lynd, Robert
It is an engaging problem in ethics whether, if you have been lent a cottage, you have the right to feed the mice.

The Peal of Bells
Chapter II (p. 9)

Poincaré, Henri
Ethics and science have their own domain which touch but do not interpenetrate. The one shows us to what goal we should aspire, the other, given the goal, teaches us how to attain it. So they can never conflict since they can never meet. There can no more be immoral science than there can be scientific morals.

The Foundations of Science
Value of Science
Introduction (p. 206)

Schweitzer, Albert
A man is truly ethical only when... he tears no leaf from a tree, plucks no flower, and takes care to crush no insects.

Philosophy of Civilization: Civilization and Ethics
Chapter XXI (p. 243)

Wilson, Edward O.
Scientists and humanists should consider together the possibility that the time has come for ethics to be removed temporarily from the hands of philosophers and biologized.

Sociobiology: The New Synthesis
Part III, Chapter 27 (p. 562)

EVIDENCE

Twain, Mark

It was not my opinion; I think there is no sense in forming an opinion when there is no evidence to form it on. If you build a person without any bones in him he may look fair enough to the eye, but he will be limber and cannot stand up; and I consider that evidence is the bones of an opinion.

Personal Recollections of Joan of Arc
Chapter II (pp. 8–9)

EVOLUTION

Barbellion, W.N.P.

How I hate the man who talks about the 'brute creation,' with an ugly emphasis on *brute*... As for me, I am proud of my close kinship with other animals. I take a jealous pride in my Simian ancestry. I like to think that I was once a magnificent hairy fellow living in the trees, and that my frame has come down through geological time via sea jelly and worms and Amphioxus, Fish, Dinosaurs, and Apes. Who would exchange these for the pallid couple in the Garden of Eden?

In Clarence Day, Jr
This Simian World (p. 2)

Bateson, William

Evolution is a process of Variation and Heredity. The older writers, though they had some vague idea that it must be so, did not study Variation and Heredity. Darwin did, and so begot not a theory, but a science.

In A.C. Seward
Darwin and Modern Science
Heredity and Variation in Modern Light (p. 88)

It is easy to imagine how Man was evolved from an Amoeba, but we cannot form a plausible guess as to how *Veronica agrestis* and *Veronica polita* were evolved, either one from the other, or both from a common form. We have not even an inkling of the steps by which a Silver Wyandotte fowl descended from *Gallus bankiva*, and we can scarcely even believe that it did.

In J. Arthur Thomsom
Concerning Evolution
Chapter II, section 11 (p. 99)

Bounoure, Louis

Evolutionism is a fairy tale for grown-ups. This theory has helped nothing in the progress of science.
It is useless.

The Advocate
March 8, 1984 (p. 17)

Bryan, William Jennings

All the ills from which America suffers can be traced back to the teaching of evolution. It would be better to destroy every other book ever written, and save just the first three verses of Genesis.

In Richard Hofstadter
Anti-Intellectualism in American Life
Chapter V (p. 125)

Carson, Rachel

It is true that I accept the theory of evolution as the most logical one that has ever been put forward to explain the development of living creatures on this earth. As far as I am concerned, however, there is absolutely no conflict between a belief in evolution and a belief in God as the creator. Believing as I do in evolution, I merely believe that is the method by which God created, and is still creating, life on earth. And it is a method so marvelously conceived that to study it in detail is to increase—and certainly never to diminish—one's reverence and awe both for the Creator and the process.

In Paul Brooks
The House of Life: Rachel Carson at Work
The Writer and His Subject (p. 9)

Darwin, Charles

... the expression often used by Mr Herbert Spencer of the Survival of the Fittest is more accurate, and is sometimes equally convenient.

The Origin of Species
Chapter III (p. 32)

When I view all beings not as special creations, but as the lineal descendants of some few beings which lived long before the first bed of the Cambrian system was deposited, they seem to me to become ennobled.

The Origin of Species
Chapter XV (p. 243)

There is grandeur in this view of life, with its several powers, having been originally breathed by the Creator into a few forms or into one; and that, whilst this planet has gone cycling on according to the fixed law of gravity, from so simple a beginning endless forms most beautiful and most wonderful have been, and are being evolved.

The Origin of Species
Chapter XV (p. 243)

On the same principle, if a man were to make a machine for some special purpose but were to use old wheels, springs and pulleys, only slightly altered, the whole machine, with all its parts, might be said to be specially contrived for its present purpose. Thus throughout nature almost every part of each living being has probably served, in a slightly modified condition, for diverse purposes, and has acted in the living machinery of many ancient and distinct specific forms.

The Works of Charles Darwin
Volume 17
The Various Contrivances by Which Orchids are Fertilized by Insects
Chapter IX (p. 283)

de Chardin, Teilhard

Is evolution a theory, a system, or a hypothesis? It is much more—it is a general postulate to which all theories, all hypotheses, all systems must henceforward bow and which they must satisfy in order to be thinkable and true. Evolution is a light which illuminates all facts, a trajectory which all lines of thought must follow—this is what evolution is.

The American Biology Teacher
In Theodosius Dobzhansky
Nothing in Biology Makes Sense Except in the Light of Evolution (p. 129)
Volume 35, Number 3, March 1973

Disraeli, Benjamin

You know, all is development. The principle is perpetually going on. First, there was nothing, then there was something; then—I forget the next—I think there were shells, then fishes; then we came—let me see—did we come next? Never mind that; we came at last. And at the next change there will be something very superior to us—something with wings. Ah! That's it: we were fishes, and I believe we shall be crows.

Tancred
Volume I
Chapter 9 (pp. 225–6)

What is the question now placed before society with the glib assurance which to me is most astonishing? That question is this: Is man an ape or an angel? I, my lord, I am on the side of the angels. I repudiate with indignation and abhorrence those new fangled theories.

Speech
Oxford Diocesan Conference
25 November 1864

Dobzhansky, Theodosius

I venture another, and perhaps equally reckless generalization—nothing makes sense in biology except in the light of evolution, *sub specie evolutionis*. If the living world has not arisen from common ancestors by means of an evolutionary process, then the fundamental unity of living things is a hoax and their diversity, a joke.

American Zoologist
Biology, Molecular and Organismic (p. 449)
Volume 4, 1964

Evolution comprises all the stages of the development of the universe: the cosmic, biological, and the human or cultural developments. Attempts to restrict the concept of evolution to biology are gratuitous. Life is a product of the evolution of inorganic nature, and man is a product of the evolution of life.

Science
Changing Man (p. 409)
Volume 155, Number 3761, 27 January 1967

Evolution is a creative process, in precisely the same sense in which composing a poem or a symphony, carving a statue, or painting a picture are creative acts...It renders possible formations of living systems that would otherwise be infinitely improbable. Nothing can be simpler and more ingenious than its mode of operation: gene constellations that fit the environment survive better and reproduce more often than those that fit less well.

Genetics of the Evolutionary Process
Chapter 12
Evolution as a Creative Process (pp. 430, 431)

Dyson, Freeman

In five billion years or less, we've evolved from some sort of primordial slime into human beings. What will happen in another ten billion years? It's just utterly impossible to conceive of ourselves changing as drastically as that over and over again, for I think all you can say is that the material form that life would take on in that kind of time scale is completely open.

In Pamela Weintraub (ed.)
The Omni Interviews
Imagine...(p. 351)

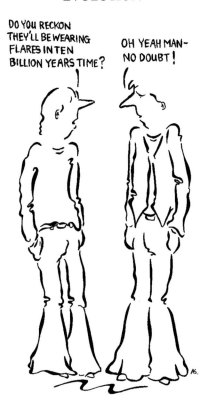

Emerson, Ralph Waldo
A SUBTLE chain of countless rings
The next unto the farthest brings;
The eye reads omens where it goes,
And speaks all languages the rose;
And, striving to be man, the worm
Mounts through all the spires of form.

Collected Poems and Translations
Nature

Futuyma, Douglas J.
Evolution, a fact rather than mere hypothesis, is the central unifying concept in biology. By extension it affects almost all other fields of knowledge and thought and must be considered one of the most influential concepts in Western thought.

Evolutionary Biology
Chapter 1 (p. 14)

Geddes, Patrick
Thomson, J. Arthur
Yet ideas of unity amid diversity, of order amid change, have also
long been growing, even finding expression, and this not merely, as
sporadically in all ages, in impressions and speculations on decline or
on better things; but in clearer and more comprehensive surveys of the
processes of change, even inquiries into its method. These, in fact, have
gone towards making up that general idea we now more or less share, of
the universe as not only orderly, but in the process of change. Changing
order, orderly change, and this everywhere—in nature inorganic and
organic, in individual and in social life—for this vast conception, now
everywhere diffusing, often expressed, rarely as yet applied, we need
some general term—and this is Evolution.

Evolution
Introduction (pp. viii–ix)

Gilbert, Sir William S.
I am, in point of fact, a particularly haughty and exclusive person, of pre-
Adamite ancestral descent. You will understand this when I tell you that I
can trace my ancestry back to a protoplasmal, primordial, atomic globule.

The Mikado
Act I (p. 349)

. . . a Darwinian Man, though well-behaved,
At best is only a monkey shaved!

Princess Ida
Act II (p. 317)

Greene, Graham
God...created a number of possibilities in case some of his prototypes failed—that is the meaning of evolution.

Travels with My Aunt
Part 2, Chapter 7 (p. 227)

Hubbard, Elbert
EVOLUTION: 1. A word that has reclassified in an entertaining manner our impermeable and eternal ignorance. 2. The growth of a thing from the simple to the complex, and the wasting away of the complex until it is simpler than ever. 3. The one superstition that is cordially hated by theologues.

The Roycroft Dictionary (p. 49)

Huxley, Julian
Evolution in the extended sense can be defined as a directional and essentially irreversible process occurring in time, which in its course gives rise to an increase of variety and an increasingly high level of organization in its products. Our present knowledge indeed forces us to the view that the whole of reality is evolution—a single process of self-transformation.

In J.R. Newman (ed.)
What Is Science
Evolution and Genetics (p. 278)

Evolutionary science is a discipline or subject in its own right. But it is the joint product of a number of separate branches of study and learning. Biology provides its central and largest component, but it has also received indispensable contributions from pure physics and chemistry, cosmogony and geology among the natural sciences, and among human studies from history and social science, archaeology and prehistory, psychology and anthropology. As a result, the present is the first period in which we have been able to grasp that the universe is a process in time and to get a first glimpse of our true relation with it. We can see ourselves as history, and can see that history in its proper relation with the history of the universe as a whole.

Evolution in Action
Chapter 1 (pp. 1–2)

There is no more need to postulate an *élan vital* or a guiding purpose to account for evolutionary progress than to account for adaptation, for degeneration or any other form of specialization...The purpose manifested in evolution, whether in adaptation, specialization, or biological progress, is only an apparent purpose. It is just as much a product of blind forces as is the falling of a stone to earth or the ebb and flow of the tides. It is we who have read purpose into evolution, as earlier men projected will and emotion into inorganic phenomena like storm or earthquake.

Evolution: The Modern Synthesis
Chapter 10
Section 3, Section 5 (pp. 568, 576)

Huxley, Thomas H.

It is very desirable to remember that evolution is not an explanation of the cosmic process, but merely a generalised statement of the method and results of that process.

Evolution & Ethics
Prolegomena (p. 6)

I have endeavored to show that no absolute line of demarcation...can be drawn between the animal world and ourselves; and I may add the expression of my belief that the attempts to draw a psychical distinction is equally futile, and even the highest faculties of feeling and of intellect begin to germinate in lower forms of life. At the same time, no one is more strongly convinced than I am of the vastness of the gulf between civilized man and the brutes; or is more certain that whether from them or not, he is assuredly not *of* them.

Collected Essays
Volume VII
On the Relations of Man to the Lower Animals

Unity of plan everywhere lies hidden under the mask of diversity of structure—the complex is everywhere evolved out of the simple.

Collected Essays
Volume III
A Lobster; or, The Study of Zoology

If the fundamental proposition of evolution is true, that the entire world, living and non-living, is the result of the mutual interaction, according to definite laws, of the forces possessed by the molecules of which the primitive nebulosity of the universe was composed, it is no less certain that the existing world lay, potentially, in the cosmic vapor, and that a sufficient intellect could, from a knowledge of the properties of the molecules of that vapor, have predicted, say the state of the Fauna of Great Britain in 1869, with as much certainty as one can say what will happen to the vapor of the breath in a cold winter's day.

In Henri Bergson
Creative Evolution
Chapter I (p. 38)

Koestler, Arthur

Evolution has made countless mistakes; for every existing species hundreds must have perished in the past; the fossil record is a wastebasket of the Chief Designer's discarded hypotheses. It is by no means unlikely that *homo sapiens*, too, is the victim of some minute error in construction—perhaps in the circuitry of his nervous system—which makes him prone to delusions, and urges him toward self-destruction.

> In Arne Tiselius and Sam Nilsson (eds)
> *The Place of Value in a World of Fact*
> The Urge to Self Destruction (p. 298)

Mayr, Ernst

The theory of evolution is quite rightly called the greatest unifying theory in biology.

> *Populations, Species and Evolution*
> Chapter 1 (p. 1)

We live in an age that places great value on molecular biology. Let me emphasize the equal importance of evolutionary biology. The very survival of man on this globe may depend on a correct understanding of the evolutionary forces and their application to man. The meaning of race, of the impact of mutation, whether spontaneous or radiation-induced, of hybridization, of competition—all these evolutionary phenomena are of the utmost importance for the human species. Fortunately the large number of biologists who continue to cultivate the evolutionary vineyard is an indication of how many biologists realize this: we must acquire an understanding of the operation of the various factors of evolution not only for the sake of understanding our universe, but indeed very directly for the sake of the future of man.

> *Cold Spring Harbor Symposia on Quantitative Biology*
> Genetics and Twentieth Century Darwinism
> Where Are We? (p. 13)
> Volume XXIV, 1959

Medawar, Peter
Medawar, J.S.

For a biologist the alternative to thinking in evolutionary terms is not to think at all.

> *The Life Science*
> Chapter 2 (p. 24)

...the testimony of Design is only for those who, secure in their beliefs already, are in no need of confirmation. This is just as well, for there is no theological comfort in the ampliation of DNA and it is no use looking to evolution. The balance sheet of evolution has so closely written a debit column of all the blood and pain that goes with the natural process that not even the smoothest accountancy can make the transaction seem morally solvent to any standards of morals that human beings are accustomed to.

The Life Science
Chapter 23 (p. 169)

Peattie, Donald Culross

In short, evolution is not so much progress as it is simply change. It does not leave all its primitive forms behind. It carries them over from age to age, well knowing that they are the precious base of the pyramid on which the more fantastic and costly experiments must be carried.

An Almanac for Moderns
April 18 (p. 31)

Roberts, Catherine

Derived from the Latin *e* (out) and *volvere* (to roll), the basic meaning of evolve is to roll out, unfold, develop. Thus, despite the seemingly random and fortuitous nature of many of the hereditary variations that permanently alter evolving individuals and populations, the scientific age generally regards the evolution of life on earth as a continuous progression from the simple to the complex and more highly organized, which has culminated in a biosphere dominated by man.

Science, Animals, and Evolution
Introduction (p. 3)

Savage, Jay M.

No serious biologist today doubts the fact of evolution...the fact of evolution is amply clear...We do not need a listing of evidences to demonstrate the fact of evolution any more than we need to demonstrate the existence of mountain ranges.

Evolution
Preface (p. v, vi)

Sherrington, Sir Charles

...Nature, often as she hugs the old, seems seldom or never to revert to a past once abandoned...Evolution can scrap but not revive.

Man on His Nature
Chapter V (p. 135)

Simpson, George Gaylord
Beck, W.S.

The fact that theories are not subject to absolute and final proof has led to a serious vulgar misapprehension. Theory is contrasted with fact as if the two had no relationship or were antitheses: 'Evolution is only a theory, not a fact.' Of course, theories are not facts. They are generalizations about facts and explanations of facts, based on and tested by facts. As such they may be just as certain—merit just as much confidence—as what are popularly termed 'facts.' Belief that the sun will rise tomorrow is the confident application of a generalization. The theory that life has evolved is founded on much more evidence than supports the generalization that the sun rises every day. In the vernacular, we are justified in calling both 'facts.'

Life: An Introduction to Biology (p. 16)

Skinner, Cornelia Otis

It is disturbing to discover in oneself these curious revelations of the validity of the Darwinian theory. If it is true that we have sprung from the ape, there are occasions when my own spring appears not to have been very far.

The Ape in Me
The Ape In Me (p. 3)

Unknown

A little boy asked his mother, "Mummy, am I descended from a monkey?" The mother replied, "I don't know, son, I never met your father's folks."

Source unknown

Watson, James D.
Evolution itself is accepted by zoologists not because it has been observed to occur or is supported by logically coherent arguments, but because it does fit all the facts of taxonomy, of palaeontology, and of geographical distribution, and because no alternative explanation is credible.

Nature
Adaptation (p. 231)
Volume 124, Number 3119, August 10, 1929

Today, evolution is an accepted fact for everyone but a fundamentalist minority, whose objections are not based on reasoning but on doctrinaire adherence to religious principles.

Molecular Biology of the Gene
Third edition
Chapter I (p. 2)

White, Timothy
You don't gradually go from being a quadruped to being a biped. What would the intermediate stage be—a triped? I've never seen one of those.

In Donald C. Johanson and Maitland A. Edey
Lucy: The Beginnings of Humankind (p. 309)

EXCEPTIONS

Bateson, William
Treasure your exceptions!...Keep them always uncovered and in sight. Exceptions are like the rough brickwork of a growing building which tells that there is more to come and shows where the next construction will be.

The Method and Scope of Genetics.
An Inaugural Lecture Delivered 23 October 1908 (p. 21)

Hackett, L.W.
Investigators are always divided into those who are looking for rules and those who are looking for exceptions.

In Marston Bates
The Natural History of Mosquitoes
Chapter XI (p. 163)

EXPERIENCE

Balfour, Arthur James
It is experience which has given us our first real knowledge of Nature and her laws. It is experience, in the shape of observation and experiment, which has given us the raw material out of which hypothesis and inference have slowly elaborated that richer conception of the material world which constitutes perhaps the chief, and certainly the most characteristic, glory of the modern mind.

The Foundations of Belief
Part II
Chapter I, Section III (p. 113)

Bierce, Ambrose
Experience, *n*. The wisdom that enables us to recognize as an undesirable old acquaintance the folly that we have already embraced.
To One who, journeying through night and fog,
Is mired neck-deep in an unwholesome bog,
Experience, like the rising of the dawn,
Reveals the path that he should not have gone.

Joel Frad Bink
The Devil's Dictionary (p. 39)

Butler, Samuel
When ninety-nine hundredths of one set of phenomena are presented while the hundredth is withdrawn without apparent cause, so that we can no longer do something which according to our past experience we shall find no difficulty whatever in doing—then we may guess what a bee must feel as it goes flying up and down a window-pane. Then we have doubts thrown upon the fundamental axiom of life—i.e. that like antecedents will be followed by like consequents. On this we go mad and die in a short time.

In Geoffrey Keynes and Brian Hill (eds)
Samuel Butler's Notebooks
Bee in a Window Pane (p. 89)

Don't learn to do, but learn in doing. Let your falls not be on a prepared ground, but let them be *bona fide* falls in the rough and tumble of the world...

In Geoffrey Keynes and Brian Hill (eds.)
Samuel Butler's Notebooks
Academicism and Myself (p. 157)

Camus, Albert

You cannot acquire experience by making experiments. You cannot create experience. You must undergo it.

Notebooks 1935–1951
Notebook I
May 1935–September 1937 (p. 5)

Coleridge, Samuel T.

To most men, experience is like the stern lights of a ship, which illumine only the track it has passed.

The Collected Works of Samuel Taylor Coleridge
Volume 14
Table Talk
Part II
September or October 1820: I 127–39 (p. 367)

da Vinci, Leonardo

Nature is full of infinite causes which were never set forth in experience.

Leonardo da Vinci's Note-Books
Book I
Life (p. 57)

Hine, Reginald L.

That is the worst of learning from experience; it takes too long. Often it takes a lifetime. 'Experience,' said Sainte-Beuve, 'is like the pole-star; it only guides a man in the evening, and rises when he is going to rest.'

Confessions of an Un-Common Attorney
Part 1
Reflections Upon the Married Estate (p. 91)

Hubbard, Elbert

EXPERIENCE: 1. The germ of power. 2. The name every one gives his mistakes. 3. Stinging and getting stung.

The Roycroft Dictionary (p. 51)

Kant, Immanuel

There can be no doubt that all our knowledge begins with experience. For how should our faculty of knowledge be awakened into action did not objects affecting our senses partly of themselves produce representations, partly arouse the activity of our understanding to compare these representations, and, by combining or separating them, work up the raw material of the sensible impressions into that knowledge of objects which is entitled experience?

Critique of Pure Reason
Introduction (p. 41)

Lacinio, Giano
...nothing short of seeing a thing will help you know it. If you wish to know that pepper is hot and that vinegar is cooling, that colocynth and absinthe are bitter, that honey is sweet, and that aconite is poison; that the magnet attracts steel, that arsenic whitens brass, and that tutia turns it of an orange color, you will, in every one of those cases, have to verify the assertion by experience.

The New Pearl of Great Price
Arguments in Favour of Our Most Glorious Art (pp. 86–7)

Mead, George H.
...in the world of immediate experience the world of things is there. Trees grow, day follows night, and death supervenes upon life. One may not say that relations here are external or even internal. They are not relations at all. They are lost in the indiscerptibility of things and events, which are what they are. The world which is the test of all observation and all scientific hypothetical reconstruction has in itself no system that can be isolated as a structure of laws, or uniformities, though all laws and formulations of uniformities must be brought to its court for its *imprimatur*.

The Philosophy of the Act
Chapter II (p. 31)

Popper, Karl
...it must be possible for an empirical scientific system to be refuted by experience.

The Logic of Scientific Discovery
Chapter I, section 6 (p. 41)

Proverb
Experience is a comb, which nature gives to men when they are bald.

In Herbert Samuel
Book of Quotations

Shapere, Dudley
One of the chief motivations behind the attempt to defend a distinction between theoretical and observational terms has been the desire to explain

how a theory can be tested against the data of experience, and how one theory can be said to "account for the facts" better than another; that is, to give a precise characterization of the idea, almost universally accepted in modern times, that the sciences are "based on experience," that they are "empirical."

Philosophical Problems of Natural Science
Introduction (p. 15)

Wilde, Oscar
Experience, the name men gave to their mistakes.

The Complete Works of Oscar Wilde
Vera
Act 2 (p. 663)

EXPERIMENT

Bohr, Neils
In every experiment on living organisms, there must remain an uncertainty as regards the physical conditions to which they are subjected, and the idea suggests itself that the minimal freedom we must allow the organism in this respect is just large enough to permit it, so to say, to hide its ultimate secrets from us.

Nature
Light and Life (p. 458)
Volume 131, Number 3309, April 1, 1933

da Vinci, Leonardo
Experiment is the interpreter of nature. Experiments never deceive. It is our judgment which sometimes deceives itself because it expects results which experiment refuses. We must consult experiment, varying the circumstances, until we have deduced general rules, for experiment alone can furnish reliable rules.

Quoted in Oswald Blackwood
Introductory College Physics (p. 47)

Darwin, Erasmus
Extravagant theories, however, in those parts of philosophy, where our knowledge is yet imperfect, are not without their use; as they encourage the execution of laborious experiments, or the investigation of ingenious deductions, to conform or refute them.

In E. Krause
Erasmus Darwin with a Preliminary Notice by Charles Darwin (pp. 139–40)

Gregg, Alan
Experiment as compared with mere observation has some of the characteristics of cross-examining nature rather than merely overhearing her.

The Furtherance of Medical Research
Chapter I (p. 7)

Kluckhohn, Clyde
Nonliterate societies represent the end results of many different experiments carried out by nature.

Mirror for Man
Chapter I (p. 15)

Whitehead, Alfred North
There is always more chance of hitting upon something valuable when you aren't too sure what you want to hit upon.

In Lucien Price
Dialogues of Alfred North Whitehead
Chapter XLII
September 11, 1945 (p. 344)

EXTINCTION

Awiakta, Marilou
To: Homo Sapiens
Re: Termination

My business is producing life.
The bottom line is
you are not cost-effective workers.
Over the millennia, I have repeatedly
clarified my management goals and objectives.
Your failure to comply is well documented.
It stems from your inability to be
a team player:
• you interact badly with co-workers
• contaminate the workplace
• sabotage the machinery
• hold up production
• consume profits
In short, you are a disloyal species.

Within the last decade
I have given you three warnings:
• made the workplace too hot for you
• shaken up your home office
• utilized plague to cut back personnel
Your failure to take appropriate action
has locked these warnings into
the Phase-Out Mode, which will result
in termination.   No appeal.

Selu: Seeking the Corn-Mother's Wisdom
Mother Nature Sends a Pink Slip (p. 88)

Carlton, J.T.
The future historians of science may well find that a crisis that was upon
us at the end of the 20th century was the extinction of the systematist,

166

the extinction of the naturalist, the extinction of the biogeographer—those who would tell the tales of the potential demise of global marine diversity.

American Zoologist
Nonextinction of Marine Invertebrates
Volume 33, 1993 (p. 507)

de Duve, Christian

The disappearance of living species is not just a blow to orchid growers, butterfly collectors, and beetle buffs. It is an irremediable loss of precious information, the biological equivalent of the burning of the library of Alexandria in 641. It is the destruction of a large part of the book of life before it can be read, the irreplaceable loss of vital clues to biological evolution and our own history. Resources of potentially great practical benefit may be lost. With each daily shrinking of the biosphere, a valuable source of food or a molecule that could have cured malaria, AIDS, or some other scourge may be vanishing forever.

Vital Dust: Life As a Cosmic Imperative
Chapter 30 (p. 275)

Editorial

Terrestrial events, like volcanic activity or change in climate or sea level, are the most immediate possible cause of mass extinctions. Astronomers should leave to astrologers the task of seeking the causes of earthly events in the stars.

The New York Times
Miscasting the Dinosaur's Horoscope
April 2, 1985

Flanders, Michael
Minale, Marcello

The Brontosaurus
Had a brain
No bigger than
A crisp;

The Dodo
Had a stammer
And the Mammoth
Had a lisp;

The Auk
Was just too Aukward—
Now they're none of them
Alive.

Each one,

(like Man),
Had shown himself
Unfit to survive.

This story
Points a moral:
Now it's
We
Who wear the pants.
The extinction
Of these species
Holds a lesson
For us
ANTS.

Creatures Great And Small...
Introductory Poem

Genesis 7:21–23
And all flesh died that moved upon the earth, both of fowl, and of cattle, and of beasts, and of every creeping thing that creepeth upon the earth... All in whose nostrils was the breath of life, of all that was in the dry land, died. And every living substance was destroyed which was upon the face of the ground... and they were destroyed from the earth.

The Bible

Hornaday, William T.
We have no right, legal, moral or commercial, to exterminate any valuable or interesting species; because none of them belong to us, to exterminate or not, as we please.

For the People of any civilized nation to permit the slaughter of the wild birds that protect its crops, its fruits and its forests from the insect hordes, is worse than folly. It is sheer orneryness and idiocy. People who are either so lazy or asinine as to permit the slaughter of their best friends deserve to have their crops destroyed and their forests ravaged.

Our Vanishing Wild Life
Chapter VI (pp. 53–4)

Muir, John
Why ought man to value himself as more than an infinitely small composing unit of the one great unit of creation?... The universe would be incomplete without man, but it would also be incomplete without the

smallest transmicroscopic creature that dwells beyond our conceitful eyes and knowledge.

A Thousand-Mile Walk to the Gulf
Chapter 6 (p. 139)

Saunders, W.E.

What good reason is there for the extermination of any form of life because it sometimes kills what we are pleased to call "game"? Are we so narrow-minded that we can undure the existence of nothing but ourselves and the things we wish to kill?

In R.J. Rutter (ed.)
W.E. Saunders—Naturalist
Saunderisms (p. 50)

Thoreau, Henry David

I love to see that Nature is so rife with life that myriads can be afforded to be sacrificed and suffered to prey on one another; that tender organizations can be so serenely squashed out of existence like pulp.

The Writings of Henry David Thoreau
Volume II
Walden (p. 350)

EXTRATERRESTRIAL LIFE

Eddington, Sir Arthur Stanley

I do not think that the whole purpose of the Creation has been staked on the one planet where we live; and in the long run we cannot deem ourselves the only race that has been or will be gifted with the mystery of consciousness. But I feel inclined to claim that *at the present time* our race is supreme; and not one of the profusion of stars in their myriad clusters looks down on scenes comparable to those which are passing beneath the rays of the sun.

The Nature of the Physical World (p. 178)

Eiseley, Loren

...nowhere in all space or on a thousand worlds will there be men to share our loneliness. There may be wisdom; there may be power; somewhere across space great instruments, handled by strange, manipulative organs, may stare vainly at our floating cloud wrack, their owners yearning as we yearn. Nevertheless, in the nature of life and in the principles of evolution we have had our answer. Of men, elsewhere, and beyond, there will be none forever.

The Immense Journey
Little Men and Flying Saucers (p. 162)

Fuller, R. Buckminster

Sometimes I think we're alone. Sometimes I think we're not. In either case, the thought is quite staggering.

In James A. Haught (ed.)
2000 Years of Disbelief (p. 290)

Jones, Harold Spencer

We see the Earth as a small planet, one member of a family of planets revolving round the Sun; the Sun, in turn, is an average star situated somewhat far out from the centre of a vast system, in which the stars are numbered by many thousands of millions; there are many millions of such systems, more or less similar to each other, peopling space to the farthest limits to which modern exploration has reached. Can it be that throughout the vast deeps of space nowhere but on our own little Earth is life to be found?

Life on Other Worlds
Chapter I (p. 19)

Metrodorus of Chios
... it would be strange if a single ear of corn grew in a large plain or there were only one world in the infinite.

In F.M. Cornford
The Classical Quarterly
Innumerable Worlds in Presocratic Philosophy (p. 13)
January 1934

Milton, John
Dream not of other Worlds; what Creatures there
Live, in what state, condition or degree...

Paradise Lost
Book VIII, l. 175–6

Oliver, Bernard M.
The last two decades have witnessed a synthesis of ideas and discoveries from previously distinct disciplines. Out of this synthesis have grown new fields of research. Thus we now have exobiology, which represents a synthesis of discoveries in astronomy, atmospheric physics, geophysics, geochemistry, chemical evolution, and biochemistry.

Project Cyclops (p. 3)

Oparin, A.I.
... there is every reason now to see in the origin of life not a "happy accident" but a completely regular phenomenon, an inherent component of the total evolutionary development of our planet. The search for life beyond Earth is thus only a part of the more general question which confronts science, of the origin of life in the universe.

In M. Calvin and O.G. Gazenko (eds)
Foundations of Space Biology and Medicine
Theoretical and Experimental Prerequisites of Exobiology
Volume I (p. 367)

Pallister, William
No one can yet show proof that there exists
A single planet save the solar ones.
But space is wide and high, and time is long,

And there are millions more of other suns.

So men imagine what they do not know
And they assume that surely there must be
Some other planets, peopled like our own;
Some other worlds with creatures such as we.

Poems of Science
Other Worlds and Ours
Life on Other Planets (p. 210)

Pope, Alexander
He, who through vast immensity can pierce,
See worlds on worlds compose one universe,
Observe how system into system runs,
What other planets circle other suns,
What varied Being peoples every star,
May tell why Heaven has made us as we are...

Alexander Pope's Collected Poems
Essay on Man
Epistle I, l. 23–8

Sagan, Carl
...there are a million other civilizations, all fabulously ugly, and all a lot smarter than us. Knowing this seems to me to be a useful and character-building experience for mankind.

In Richard Berendzen
Life Beyond Earth & the Mind of Man
Sagan (p. 64)

Shakespeare, William
Glendower: I can call spirits from the vasty deep.
Hotspur: Why, so can I, or so can any man;
        But will they come when you do call them?

The First Part of King Henry the Fourth
Act III, Scene I, l. 53–5

Horatio: O day and night, but this is wondrous strange!
Hamlet: And therefore as a stranger give it welcome.
        There are more things in heaven and earth, Horatio,
        Than are dreamt of in your philosophy.

Hamlet, Prince of Denmark
Act I, Scene V, l. 164–7

FACT

American Museum of Natural History
Every specimen is a permanent fact.

Plaque at entrance to the Earth History Hall

Beaumont, William
My opinions may be doubted, denied, or approved, according as they conflict or agree with the opinions of each individual who may read them; but their worth will be best determined by the foundation on which they rest—the incontrovertible facts.

William Beaumont: A Pioneer American Physiologist
Experiments and Observations on the Gastric Juice
and the Physiology of Digestion
Preface (p. 200)

Bernard, Claude
When we meet a fact which contradicts a prevailing theory, we must accept the fact and abandon the theory, even when the theory is supported by great names and generally accepted.

Introduction to the Study of Experimental Medicine
Part III, Chapter II (p. 164)

Bradford, Gamaliel
Observed facts must be built up, woven together, ordered, arranged, systematized into conclusions and theories by reflection and reason, if they are to have full bearing on life and the universe. Knowledge is the accumulation of facts. Wisdom is the establishment of relations. And just because the latter process is delicate and perilous, it is all the more delightful. The lofty scorn of the true philosopher for mere perception is well shown in Royer Collard's remark: 'There is nothing so despicable as a fact.' Which does not prevent philosophers or any one else from making facts the essential basis of all discussion of relations.

Darwin
Chapter II (p. 44)

...the mere collection of facts, without some basis of theory for guidance and elucidation, is foolish and profitless.

Darwin
Chapter II (p. 47)

Carpenter, William

Were we able to ascertain *facts* regarding the changes which take place in the interior of the living body as easily as the astronomer observes the place of a planet, or the chemist the decomposition of a salt, there is no reason whatever to prevent these facts being generalized in the same manner and to the same degree with those of the physical sciences.

British and Foreign Medical Review
Review of A History of the Inductive Sciences
Physiology An Inductive Science (p. 340)
Volume 5, 1838

Collins, Wilkie

"Facts?" he repeated. "Take a drop more grog, Mr Franklin, and you'll get over the weakness of believing facts! Foul play, Sir."

The Moonstone
Second Narrative
Chapter IV (p. 275)

Darwin, Charles

...no one has a right to speculate without distinct facts...

The Voyage of the Beagle
Chapter XXI (p. 378)

False facts are highly injurious to the progress of science, for they often endure long; but false views, if supported by some evidence, do little harm, for every one takes a salutary pleasure in proving their falseness: and when this is done, one path towards error is closed and the road to truth is often at the same time opened.

The Descent of Man
Part III, Chapter XXI (p. 590)

Doyle, Sir Arthur Conan

Some facts should be suppressed, or at least, a just sense of proportion should be observed in treating them. The only point in the case which deserved mention was the curious analytical reasoning from effects to causes, by which I succeeded in unravelling it.

The Complete Sherlock Holmes
The Sign of Four
Chapter 1 (p. 90)

Emerson, Ralph Waldo
To the wise, therefore, a fact is true poetry, and the most beautiful of fables.

The Collected Works of Ralph Waldo Emerson
Volume I
Nature
Prospects (p. 44)

Gooday, Graeme
We have tables properly arranged in regard to light, microscopes and dissecting instruments, and we work through the structure of a certain number of animals and plants... the student has before him, first, a picture of the structure he ought to see; secondly the structure itself worked out, and if with these aids, and such needful explanations and practical hints as a demonstrator can supply, he cannot make out the facts for himself in the material supplied to him, he had better take to some other pursuit than that of biological science.

British Journal for the History of Science
Nature in the Laboratory (pp. 339-40)
Volume 24, 1991

Gould, Stephen Jay
Facts cannot be divorced from cultural contexts.

Natural History
What Color is a Zebra? (p. 16)
Volume 90, Number 8, August 1981

Huxley, Thomas H.
Spencer's idea of a tragedy is a deduction killed by a fact.

In William Irvine
Apes, Angels, and Victorians
Chapter III (p. 30)

Sit down before fact as a little child... follow humbly and to whatever abysses Nature leads, or you shall learn nothing.

In Leonard Huxley (ed.)
Life and Letters of Thomas Henry Huxley
Volume I
Huxley to Kingsley
September 23, 1860 (p. 316)

James, William
'Facts' are the bounds of human knowledge, set for it, not by it.

The Will to Believe and Other Popular Essays in Popular Philosophy
On Some Hegelisms (p. 202)

Jeffreys, H.
There are some current 'theories' that, when divested of begged questions, reduce to the non-controversial statement, 'Here are some facts and there may be some relation between them.'

Theory of Probability
Chapter VIII, Section 8.5 (p. 419)

Latour, Bruno
...a fact is what is collectively stabilised from the midst of controversies when the activity of later papers does not consist only of criticism or deformation but also of confirmation.

Science in Action
Literature (p. 42)

Lavoisier, Antoine
We must trust to nothing but facts: these are presented to us by nature, and cannot deceive. We ought, in every instance, to submit our reasoning to the test of experiment and never to search for truth but by the natural road of experiment and observation.

Elements of Chemistry
Preface (p. 2)

Lukasiewicz, J.
Facts whose effects have disappeared altogether, and which even an omniscient mind could not infer from those now occurring, belong to the realm of possibility. One cannot say about them that they took place, but only that they were *possible*.

In L. Borkowski (ed.)
Selected Works
On Determinism (p. 128)

Millay, Edna St Vincent
Upon this gifted age, in its dark hour,
Rains from the sky a meteoric shower
Of facts... they lie unquestioned, uncombined,
Wisdom enough to leach us of our ill
Is daily spun; but there exists no loom
To weave it into fabric;...

Collected Sonnets
Three Sonnets in Tetrameter
Sonnet III (p. 140)

Shaw, George Bernard
Facts mean nothing by themselves. All the people at present crowding the Strand are facts. Nobody can possibly know the facts. Naturalists *collect*

a few. Men of genius *select* a fewer few, and lo! a drama or a hypothesis. Genius is a sense of values and significances (the same thing). Without this sense facts are useless mentally. With it a Goethe can do more with ten facts than an encyclopedia compiler with ten thousand.

In J. Percy Smith (ed.)
Selected Correspondence of Bernard Shaw: Bernard Shaw to H.G. Wells
Letter to H. G. Wells (pp. 152–3)
2 August 1929

Tansley, A.G.
We must never conceal from ourselves that our concepts are creations of the human mind which we impose on the facts of nature, that they are derived from incomplete knowledge, and therefore will never *exactly* fit the facts, and will require constant revision as knowledge increases.

Journal of Ecology
The Classification of Vegetation and the Concept of Development (p. 120)
Volume 8, Number 2, June 1920

Thoreau, Henry David
Let us not underrate the value of a fact; it will one day flower in a truth. It is astonishing how few facts of importance are added in a century to the natural history of any animal. The natural history of man himself is still being gradually written.

The Writings of Henry David Thoreau
Volume V
Natural History of Massachussetts (p. 130)

Wilde, Oscar
Facts are not merely finding a footing-place in history, but they are usurping the domain of Fancy, and having invaded the kingdom of Romance. Their chilling touch is over everything. They are vulgarising mankind.

Intentions
The Decay of Lying (p. 27)

FIBONACCI

Kauffman, Stuart

Pick up a pinecone and count the spiral rows of scales. You may find eight spirals winding up to the left and 13 spirals winding up to the right, or 13 left and 21 right spirals, or other pairs of numbers. The striking fact is that these pairs of numbers are adjacent numbers in the famous Fibonacci series: 1, 1, 2, 3, 5, 8, 13, 21... Here, each term is the sum of the previous two terms. The phenomenon is well known and called phyllotaxis. Many are the efforts of biologists to understand why pinecones, sunflowers, and many other plants exhibit this remarkable pattern. Organisms do the strangest things, but all these odd things need not reflect selection or historical accident. Some of the best efforts to understand phyllotaxis appeal to a form of self-organization. Paul Green, at Stanford, has argued persuasively that the Fibonacci series is just what one would expects as the simplest self-repeating pattern that can be generated by the particular growth processes in the growing tips of the tissues that form sunflowers, pinecones, and so forth. Like a snowflake and its sixfold symmetry, the pinecone and its phyllotaxis may be part of order for free.

At Home in the Universe
Chapter 8 (p. 151)

FISH

Hemingway, Ernest
He is a great fish and I must convince him, he thought. I must never let him learn his strength nor what he could do if he made his run...but thank God, they are not as intelligent as we who kill them; although they are more noble and more able.

The Old Man and the Sea (p. 61)

Hunt, Leigh
You strange, astonish'd-looking, angle-faced,
Dreary-mouth'd, gaping wretches of the sea,
Gulping salt-water everlastingly,
Cold-blooded, though with red your blood be graced,
And mute, through dwellers in the roaring waste;
And you, all shapes beside, that fishy be—
Some round, some flat, some long. all devilry,
Legless, unloving, infamously chaste:—

The Poetical Works of Leigh Hunt
The Fish, The Man, and the Spirit (p. 250)

MacLeish, Archibald
Plunge beneath the ledge of coral
Where the silt of sunlight drifts
Like dust that settles toward a floor—
As slow as that: feel the lifting
Surge that rustles white above
But here is only movement deep
As breathing: watch the reef fish hover
Dancing in their silver sleep
Around their stone, enchanted tree:...

The Collected Poems of Archibald MacLeish
The Reef Fisher

Smith, Langdon
When you were a tadpole and I was a fish,
In the Paleozoic time.

In E. Halderman-Julius
Poems of Evolution
Evolution

Pallister, William
Fifteen thousands of species of FISHES are known,
And some kinds are enormous and others minute;
They are widespread, wherever their tribes can be grown
And all seeking the foods which their habits will suit;
Some migrating in millions that their spawn may be sown,
Some in depths of the ocean, but rarely alone.

Poems of Science
Beginnings
Animal Life (p. 140)

Peacock, Thomas Love
Premising that this is a remarkably fine slice of salmon, there is much to
be said about fish: but not in the way of misnomers. Their names are
single and simple. Perch, sole, cod, eel, carp, char, skate, tench, trout, brill,
bream, pike, and many others, plain monosyllables: salmon, dory, turbot,
gudgeon, lobster, whitebait, grayling, haddock, mullet, herring, oyster,
sturgeon, flounder, turtle, plain disyllables: only two trisyllables worth
naming: anchovy and mackerel; unless any one should be disposed to
stand up for halibut, which, for my part, I have excommunicated.

Gryll Grange
Misnomers (p. 12)

Thoreau, Henry David
Who hears the fish when they cry?

The Writings of Henry David Thoreau
Volume I
A Week on the Concord and Merrimack Rivers
Saturday (p. 45)

BARRACUDA

Gardner, John
Slowly, slowly, he cruises,
And slowly, slowly, he chooses
Which kind of fish he prefers to take this morning;
Then without warning

The Barracuda opens his jaws, teeth flashing,
And with a horrible, horrible grinding and gnashing,
Devours a hundred poor creatures and feels no remorse.
…
"But," (as he says
With an evil grin)
"It's actually not my fault, you see:
I've nothing to do with the tragedy;
I open my mouth for a yawn and—ah me—
They all
    swim
        in."

A Child's Beastiary
The Barracuda

CODFISH

Unknown
The codfish lays a thousand eggs
The homely hen lays one.
The codfish never cackles
To tell you what she's done.
And so we scorn the codfish
While the humble hen we prize
Which only goes to show you
That it pays to advertise.

In Mark Kurlansky
Cod (p. 29)

GUPPY

Nash, Ogden
Whales have calves,
Cats have kittens,
Bears have cubs,
Bats have bittens.
Swans have cygnets,
Seals have puppies,
But guppies just have little guppies.

Verses from 1929 On
The Guppy

HERRING

Cuppy, Will
Some fishes become extinct, but Herrings go on forever. Herrings spawn at all times and places and nothing will induce them to change their ways. They have no fish control. Herrings congregate in schools, where they learn nothing at all. They move in vast numbers in May and October. Herrings subsist upon Copepods and Copepods subsist upon Diatoms and Diatoms just float around and reproduce. Young Herrings or Sperling or Whitebait are rather cute. They have serrated abdomens. The skull of the Common or Coney Island Herring is triangular, but he would be just the same anyway. (The nervous system of the Herring is fairly simple. When the Herring runs into something the stimulus is flashed to the forebrain, with or without results.)

How to Become Extinct
The Herring (p. 13)

KIPPER

Nash, Ogden
For half a century, man and nipper,
I've doted on a tasty kipper,
But since I am no Jack the Ripper
I wish the kipper had a zipper.

Everyone But Thee and Me
The Kipper (p. 63)

PICKEREL

Thoreau, Henry David
The swiftest, wariest, and most ravenous of fishes.

The Writings of Henry David Thoreau
Volume I
A Week on the Concord and Merrimack Rivers (p. 29)

SALMON

McGregor, James
Oh! For the thrill of a Highland stream,
With the bending rod of a fisherman's dream,
The screaming reel and flying line,
Where the far-flung pearl-drops wetly shine—
The sudden leap, then the silent strife,

While the salmon grimly fights for life;
As a worthy foe, or a regal dish,
We respect this gallant fighting fish.

In Arnold Silcock
Verse and Worse
Ode to a Salmon (p. 21)

SCULPIN

Holmes, Oliver Wendell
Now the Sculpin (*Cottus Virginianus*) is a little water beast which pretends to consider itself a fish, and, under that pretext, hangs about the piles on which West Boston Bridge is built, swallowing the bait and hook intended for flounders. On being drawn from the water, it exposes an immense head, a dimunitive bony carcass, and a surface so full of spines, ridges, ruffles and frills that the naturalist have not been able to count them without quarreling about their number.

More Yankee Drolleries
The Professor at the Breakfast Table
Chapter I (pp. 1–2)

SEA HORSE

Kraus, Jack
SEA HORSE: Philly of flounder.

Quote
October 23, 1966 (p. 17)

SHARK

Nash, Ogden
How many Scientists have written
The shark is gentle as a kitten!
Yet this I know about the shark:
His bite is worser than his bark.

Verses from 1929 On
The Shark

WOOF!

SMELT

Nash, Ogden
Oh, why does man pursue the smelt?
It has no valuable pelt,
It boasts of no escutcheon royal,
It yields not ivory or oil,
Its life is dull, its death is tame,
A fish as humble as its name.
Yet—take this salmon somewhere else,
And bring me half a dozen smelts.

Verses from 1929 On
The Smelt

STURGEON

Longfellow, Henry Wadsworth
On the white sand of the bottom
Lay the monster Mishe-Nahma,
Lay the sturgeon, King of Fishes;
Through his gills he breathed the water,
With his fins he fanned and winnowed,

With his tail he swept the sand-floor.
    There he lay in all his armor;
On each side a shield to guard him,
Plates of bone upon his forehead,
Down his sides and back and shoulders
Plates of bone with spines projecting!

The Complete Writings of Henry Wadsworth Longfellow
Volume II
Hiawatha
Part VIII

WHITING

Carroll, Lewis
"Will you walk a little faster?" said a whiting to a snail,
"There's a porpoise close behind us, and he's treading on my tail."

The Complete Works of Lewis Carroll
Alice's Adventures in Wonderland
Chapter X (p. 107)

FLOWERS

Beecher, Henry Ward
He who only does not appreciate floral beauty is to be pitied like any other man who is born imperfect.

Star Papers
A Discourse on Flowers (p. 94)

Flowers have an expression of countenance as much as men or animals. Some seem to smile; some have a sad expression; some are pensive and diffident; others again are plain, honest and upright, like the broad-faced sunflower and the hollyhock.

Star Papers
A Discourse on Flowers (p. 100)

Flowers are the sweetest things God ever made and forgot to put a soul into.

Life Thoughts (p. 234)

Blake, William
To create a little flower is the labor of ages.

The Complete Poetry and Prose of William Blake
The Marriage of Heaven and Hell
Proverbs of Hell, l. 56

Child, Lydia M.
Flowers have spoken to me more than I can tell in written words. They are the hieroglyphics of angels, loved by all men for the beauty of the character, though few can decypher even fragments of their meaning.

Letters from New York
Letter XXVI
September 1, 1842

Millay, Edna St Vincent
I will be the gladdest thing
Under the sun!
I will touch a hundred flowers
And not pick one.

Collected Poems
Afternoon on a Hill (p. 33)

Milton, John
Into the blissful field, through Groves of Myrrhe,
And flouring Odours, Cassia, Nard, and Balme;
A Wilderness of sweets...

Paradise Lost
Book V, l. 294–7

...the bright consummate floure...

Paradise Lost
Book V, l. 481

Rossetti, Christina G.
Flowers preach to us if we will hear.

The Complete Poems of Christina Rossetti
Volume I
Consider the Lilies of the Field (p. 76)

Unknown
The pistol of a flower is its only protection against insects.

Source unknown

von Frisch, Karl
One can see that the colors of the flowers have been developed as an adaptation to the color sense of their visitors. It is evident that they are not designed for the human eye. But this should not prevent us from delighting in their beauty.

Bees: Their Vision, Chemical Senses, and Language
The Color Sense of Bees (p. 13)

AMARANTH

Browning, Elizabeth Barrett
Nosegays! leave them for the waking;
Throw them earthward where they grew;
Dim are such beside the breaking
Amaranths he looks unto:

Folded eyes see brighter colors than the open ever do.

The Complete Poetical Works of Elizabeth Barrett Browning
A Child Asleep
Stanza II

Milton, John
Bid *Amaranthus* all his beauty shed,
And Daffodillies fill their cups with tears,
To strew the Laureate Hearse where *Lycid* lies.

Lycidas
l. 149–51

Immortal Amarant, a Flour which once
In Paradise, fast by the Tree of Life,
Began to bloom, but soon for man's offence,
To Heav'n remov'd, where first it grew, there grows,
And flours aloft shading the Fount of Life.

Paradise Lost
Book III, l. 353–7

Moore, Thomas
Amaranths, such as crown the maids
That wander through Zamara's shades...

The Poetical Works of Thomas Moore
Lalla Rookh
Light of the Harem (p. 258)

AMARYLLIS

Tennyson, Alfred
Where, here and there, on sandy beaches
A milky-bell'd amaryllis blew!

The Complete Poetical Works of Tennyson
The Daisy, Stanza 4

ANEMONE

Bryant, William Cullen
Within the woods,
Whose young and half transparent leaves scarce cast
A shade, gray circles of anemones
Danced on their stalks;...

Poems
The Old Man's Counsel

Goodale, Elaine
Thy subtle charm is strangely given,
My fancy will not let thee be,—
Then poise not thus 'twixt earth and heaven,
O white anemone!

All Round the Year
Anemone
Stanza 6

Moore, Thomas
Anemones and Seas of Gold,
And new-blown lilies of the river,
And those sweet flow'rets that unfold
Their buds on Camadera's quiver...

The Poetical Works of Thomas Moore
Lalla Rookh
Light of the Harem (p. 258)

AQUILEGIA

Taylor, Bayard
The aquilegia sprinkled on the rocks
A scarlet rain; the yellow violets
Sat in the chariot of its leaves; the phlox
Held spikes of purple flame in meadows wet,
And all the streams with vernal-scented reed
Were fringed, and streaky bells of miskodeed.

The Poetical Works of Bayard Taylor
Mon-Da-Min, Stanza 17

ARBUTUS, TRAILING

Terry, Rose
Darlings of the forest!
Blossoming alone
When Earth's grief is sorest
For her jewels gone—
Ere the last snow-drift melts, your tender buds have blown.

Poems
Trailing Arbutus

ASPHODEL

Browning, Elizabeth Barrett
With her ankles sunken in asphodel
She wept for the roses of earth which fell...

The Complete Poetical Works of Elizabeth Barrett Browning
Calls on the Heart
Stanza IV

Pope, Alexander
By the streams that ever flow,
By the fragrant winds that blow
O'er the Elysian flow'rs;
By those happy souls who dwell
In yellow mead of asphodel.

Alexander Pope's Collected Poems
Ode on St Celia's Day
Stanza V, l. 71–5

ASTER

Emerson, Ralph Waldo
Chide me not, laborious band,
For the idle flowers I brought;
Every aster in my hand
Goes home loaded with a thought.

Collected Poems and Translations
The Apology

Whitman, Sarah Helen
And still the aster greets us as we pass
With her faint smile,—among the withered grass
Beside the way, lingering as loth of heart,
Like me, from these sweet solitudes to part.

Poems
A Day of the Indian Summer

AZALEA

Goodale, Dora Read
O far away in yonder leafy copse
The wandering thrush has flown,
And close along the wooded steep

We know an influence passing deep,
The Summer light, The Summer tone,
The rare azalea makes her own,—
And we are not alone

All Round the Year
Wild Azalea

BARBERRIES

Aldrich, Thomas Bailey
In scarlet clusters o'er the grey stone-wall
The barberries lean in thin autumnal air:
Just when the fields and garden-plots are bare,
And ere the green leaf takes the tint of fall,
They come to make the eye a festival!
Along the road, for miles, their torches flare.

The Poems of Thomas Bailey Aldrich
Sonnets
Barberries

BASIL

Moore, Thomas
...the basil tuft, that waves
Its fragrant blossom over graves...

The Poetical Works of Thomas Moore
Lalla Rookh
Light of the Harem (p. 259)

BEAN

Ingelow, Jean
I know the scent of bean-fields.

Poems
Gladys and Her Island

BLOODROOT

Bryant, William Cullen
Of Sanguinaria, from whose brittle stem
The red drops fell like blood.

Poems
The Fountain

Goodale, Elaine
O bloodroot! in thy tingling veins
The sap of life runs cold and clear;
I break thy shining stem, and fear
No conscious guilt, no lasting stains.

All Round the Year
Bloodroot

BORAGE

MacDonald, George
The flaming rose gloomed swarthy red;
The borage gleams more blue;
Dim starred with white flowers, a flowering bed
Glimmer the rich dusk through.

The Poetical Works of George MacDonald
Songs of the Summer Night
Part III

BRAMBLE

Chaucer, Geoffrey
But many a maiden, bright in bowser
Did long for him for paramour
When they were best asleep;
But chaste he was, no lecher sure,
And sweet as is the bramble-flower
That bears a rich red hepe.

The Canterbury Tales
Sir Thopas (p. 397)

Elliott, Ebenezer
Thy fruit full well the schoolboy knows,
Wild brambles of the brake!
So, put thou forth thy small white rose;

I love it for his sake.

The Poetical Works of Ebenezer Elliott
To the Bramble Flower

BUTTERCUP

Browning, Elizabeth Barrett
He likes the poor things of the world the best,
I would not, therefore, if I could be rich.
It pleases him to stoop for buttercups.

The Complete Poetical Works of Elizabeth Barrett Browning
Aurora Leigh
Book IV
l. 210–12

Mulock, Dinah Maria
The buttercups across the field
Made sunshine rifts of splendor...

Miss Mulock's Poems
A Silly Song

CAMOMILE

Shakespeare, William
... for though the camomile, the more it is trodden on the faster it grows...

The First Part of King Henry the Fourth
Act II, Scene IV, l. 438–9

CARDINAL FLOWER

Goodale, Dora Read
Whence is yonder flower so strangely bright?
Would the sunset's last reflected shine
Flame so red from that dead flush of light?
Dark with passion is its lifted line,
Hot, alive, amid the falling night.

All Round the Year
Cardinal Flower

CARNATION

Milton, John
Each Flour of slender stalk, whose head though gay
Carnation, Purple, Azure, or spect with Gold hung drooping unsustained...

Paradise Lost
Book IX, l. 429

CASSIA

Ingelow, Jean
While cassias blossom in the zone of calms.

Poems
Sand Martins

CATALPA

Bryant, William Cullen
...the Catalpa's blossoms flew,
Light blossoms, dropping on the grass like snow.

Poems
The Winds
Stanza I

CELANDINE

Wordsworth, William
Long as there's a sun that sets,
Primroses will have their glory;
Long as there are violets,
They will have a place in story:
There's a flower that shall be mine,
'Tis the little Celandine.

The Complete Poetical Works of William Wordsworth
To the Small Celandine
Stanza I

CHAMPAC

Moore, Thomas
The maid of India, blessed again to hold
In her full lap the Champac's leaves of gold.

The Poetical Works of Thomas Moore
Lalla Rookh
The Veiled Prophet of Khorassan

CHRYSANTHEMUM

Wilde, Oscar
Chrysanthemums from gilded argosy
Unload their gaudy scentless merchandise.

Poems
Humanitad, Stanza 11

CLEMATIS

Goodale, Dora Read
Where the woodland streamlets flow,
Gushing down a rocky bed,
Where the tasseled alders grow,
Lightly meeting overhead,
When the fullest August days
Give the richness that they know,
Then the wild clematis comes,
With her wealth of tangled blooms,
Reaching up and drooping low.

All Round the Year
Wild Clematis

COLUMBINE

Bryant, William Cullen
Or columbines, in purple dressed,
Nod o'er the ground-bird's hidden nest.

Poems
To the Fringed Gentian
Stanza II

Ingelow, Jean
O columbine, open your folded wrapper,

Where two twin turtle-doves dwell!
O cuckoopint, toll me the purple clapper
That hangs in your clear green bell!

Songs of Seven
Seven Times One

Rusby, Henry H.
Sweet flower of the golden horn,
Thy beauty passeth praise!
But why should spring thy gold adorn
Most meet for summer days?

To the Golden Columbine
Source unknown

COMPASS-PLANT

Longfellow, Henry Wadsworth
Look at this vigorous plant that lifts its head from the meadow,
See how its leaves are turned to the north, as true as the magnet;
This is the compass-flower, that the finger of God has planted...

The Complete Writings of Henry Wadsworth Longfellow
Volume II
Evangeline
Part II, Stanza IV, l. 140

CORAL-TREE

Moore, Thomas
The crimson blossoms of the coral-tree
In the warm isles of India's sunny sea.

The Poetical Works of Thomas Moore
Lalla Rookh
The Veiled Prophet of Khorassan

COWSLIP

Hood, Thomas
The cowslip is a country wench.

The Poetical Works of Thomas Hood
Flowers
Stanza I

DAFFODIL

Tennyson, Alfred
When the face of night is fair on the dewy downs,
And the shining daffodil dies...

The Complete Poetical Works of Tennyson
Maud
Part III, Stanza 1

Wordsworth, William
I wandered lonely as a cloud
That floats on high o'er vales and hills,
When all at once I saw a crowd,
A host, of golden daffodils...

The Complete Poetical Works of William Wordsworth
I Wandered Lonely as a Cloud

DAHLIA

Elliott, Ebenezer
The Vicar's house is smother'd in its roses,
His garden glows with dahlias large and new.

The Poetical Works of Ebenezer Elliott
The Vicarage

DAISY, OX-EYE

Goodale, Dora Read
Clear and simple in white and gold,
Meadow blossom, of sunlit spaces,—
The field is full as it well can hold
And white with the drift of the ox-eye daisies!

All Round the Year
Daisies

DANDELION

Beecher, Henry Ward
You can not forget, if you would, those golden kisses all over the cheeks
of the meadow, queerly called dandelions.

Star Papers
A Discourse of Flowers (p. 97)

DODDER

Davis, Sarah F.
In the roadside thicket hiding,
Sing, robin, sing!
See the yellow dodder, gliding,
Ring, Bluebells, ring!
Like a living skein inlacing,
Coiling, climbing, turning, chasing,
Through the fragrant sweet-fern racing—
Laugh, O murmuring Spring!

Summer Song
Source unknown

FERN

Burroughs, John
I know of nothing in vegetable nature that seems so really to be *born* as the ferns. They emerge from the ground rolled up, with a rudimentary and "touch-me-not" look, and appear to need a maternal tongue to lick them into shape, The sun plays the wet-nurse to them, and very soon they are out of that uncanny covering in which they come swathed and take their places with other green things.

Signs and Seasons
A Spring Relish (p. 193)

FLAG

Shelley, Percy Bysshe
And nearer to the river's trembling edge
There grew broad flag-flowers, purple
    prankt with white;
And starry river buds among the sedge;
And floating water-lilies broad and bright.

Shelley: Selected Poetry, Prose and Letters
The Question
Stanza IV

FLOWER-DE-LUCE

Longfellow, Henry Wadsworth
O flower-de-luce, bloom on, and let the river
Linger to kiss thy feet!

O flower of song, bloom on, and make forever
The world more fair and sweet.

The Complete Writings of Henry Wadsworth Longfellow
Volume III
Flower-de-luce
Stanza 8

FORGET-ME-NOT

Coleridge, Samuel T.
The blue and bright-eyed floweret of the brook,
Hope's gentle gem, the sweet Forget-me-not!

The Complete Poetical Works of Samuel Taylor Coleridge
Volume I
The Keepsake, l. 12–13

FOXGLOVE

Ingelow, Jean
An empty sky, a world of heather,
Purple of foxglove, yellow of broom;
We two among them wading together,
Shaking out honey, treading perfume.

Poems
Divided
Stanza I

FURZE

Goldsmith, Oliver
With blossom'd furze unprofitably gay.

The Complete Poetical Works of Oliver Goldsmith
The Deserted Village, l. 194

GENTIAN

Bryant, William Cullen
And the blue gentian-flower, that, in the breeze,
Nods lonely, of her beauteous race the last.

Poems
November

GILLYFLOWER

Shakespeare, William
The fairest flowers o'the season
Are our carnations and streak'd gillyvors,
Which some call nature's bastards.

The Winter's Tale
Act IV, Scene III, l. 81–3

GOLDENROD

Jackson, Helen Hunt
I know the lands are lit
With all the autumn blaze of Golden Rod...

Verses
Asters and Goldenrod

GORSE

Ingelow, Jean
But I have seen
The gay gorse bushes in their flowering time.

Poems
Gladys and her Island

HAREBELL

Shakespeare, William
Thou shalt not lack
The flower that's like thy face, pale primrose nor
The azured harebell, like thy veins...

Cymbeline
Act IV, Scene II, l. 220–2

HEATH

Pope, Alexander
E'en wild heath displays her purple dyes,
And 'midst the desert fruitful fields arise.

Alexander Pope's Collected Poems
Windsor Forest, l. 25–6

HELIOTROPE

Stedman, E.C.
O sweetest of all the flowrets
That bloom where angels tread!
But never such marvelous odor
From heliotrope was shed...

The Poetical Works of Edmund Clarence Stedman
Heliotrope
Stanza 2

HEPATICA

Goodale, Dora Read
All the woodland path is broken
By warm tints along the way,
And the low and sunny slope
Is alive with sudden hope
When there comes the silent token
Of an April day,—
Blue hepatica!

All Round the Year
Hepatica

HOLLYHOCK

Ingelow, Jean
...and Queen hollyhocks,
With butterflies for crowns...

Poems
Honors
Part I, Stanza 5

HONEYSUCKLE

Landon, L.E.
...and scarce a beech
Was there with a honeysuckle link'd
Around, with its red tendrils and pink flowers...

The Poetical Works of Miss Landon
Volume I
The Oak

Tennyson, Alfred
The honeysuckle round the porch has woven its wavy bowers...

The Complete Poetical Works of Tennyson
The May Queen
Stanza 8

HYACINTH

Montgomery, James
Here Hyacinths of heavenly blue
Shook their rich tresses to the morn.

Poetical Works of James Montgomery
Volume II
The Adventure of a Star

Shelley, Percy Bysshe
And the hyacinth purple, and white, and blue,
Which flung from its bells a sweet peal anew
Of music so delicate, soft and intense,
It was felt like an odour within the sense.

Shelley: Selected Poetry, Prose and Letters
The Sensitive Plant
Part I, Stanza 7

IVY

Bailey, Philip James
For ivy climbs the crumbling hall
To decorate decay.

Festus: a Poem
Party and Entertainment (p. 272)

Browning, Elizabeth Barrett
...walls must get the weather stain
Before they grow the ivy!

The Complete Poetical Works of Elizabeth Barrett Browning
Aurora Leigh
Book VIII, l. 694–5

JASMINE

Bryant, William Cullen
And at my silent window-sill
The jessamine peeps in.

Poems
The Hunter's Serenade

LILY

Aldrich, Thomas Bailey
I like not lady-slippers,
Not yet the sweet-pea blossoms,
Nor yet the flaky roses,
Red or white as snow;
I like the chaliced lilies,
The heavy Eastern lilies,
The gorgeous tiger-lilies,
That in our garden grow.

The Poems of Thomas Bailey Aldrich
Tiger Lilies
Stanza 1

Cook, Eliza
The citron-tree or spicy grove for me would never yield
A perfume half so grateful as the lilies of the field.

The Poetical Works of Eliza Cook
England
Stanza 2

LILY-OF-THE-VALLEY

Croly, George
White bud, that in meek beauty so dost lean
Thy cloistered cheek as pale as moonlight snow,
Thou seem'st beneath thy huge, high leaf of green,
An Eremite beneath his mountain's brow.

The Poetical Works of the Rev. George Croly
The Lily of the Valley

LOTUS

Heine, Heinrich
The lotus flower is troubled
At the sun's resplendent light;
With sunken head and sadly
She dreamily waits for the night.

Book of Songs
Lyrical Interlude
Number 10 (p. 52)

LOVE LIES BLEEDING

Swinburne, Algernon Charles
Loves lies bleeding in the bed whereover
Roses lean with smiling mouths or pleading;
Earth lies laughing where the sun's dart clove her:
Love lies bleeding.

The Complete Poetical Works of Algernon Charles Swinburne
Poetical Works
Volume I
Love Lies Bleeding

MANOLIA GRANDIFLORA

Cranch, C.P.
Majestic flower! How purely beautiful
Thou art, as rising from thy bower of green,
Those dark and glossy leaves so thick and full,
Thou standest like a high-born forest queen
Among thy maidens clustering round so fair…
I breathe the perfume, delicate and strong,
That comes like incense from thy petal-bower;
My fancy roams those southern woods along,
Beneath that glorious tree, where deep among
The unsunned leaves thy large white flower-cups hung!

Collected Poems of Christopher Pearse Cranch
Poem to the Magnolia Grandiflora

MARIGOLD

Keats, John
Open afresh your round of starry folds,

Ye ardent marigolds!
Dry up the moisture from your golden lips.

Complete Poems
I Stood Tiptoe Upon a Little Hill, l. 47–9

MARSH MARIGOLD

Swinburne, Algernon Charles
A little marsh-plant, yellow green,
And pricked at lip with tender red,
Tread close, and either way you tread,
Some faint black water jets between
Least you should bruise the curious head.

The Complete Poetical Works of Algernon Charles Swinburne
Poetical Works
Volume I
The Sundew

MEADOW RUE

Goodale, Elaine
When emerald slopes are drowned in song,
When weary grows the unclouded blue,
When warm winds sink in billowy bloom,
And flood you with a faint perfume,
One moment leaves the rapturous throng
To seek the haunts of meadow rue!

All Round the Year
Meadow Rue
Stanza 4

MIMOSA

Shelley, Percy Bysshe
A Sensitive Plant in a garden grew,
And the young winds fed it with silver dew,
And it opened its fan-like leaves to the light,
And clothed them beneath the kisses of night.

Shelley: Selected Poetry, Prose and Letters
The Sensitive Plant
Part I, Stanza 1

MOCCASIN FLOWER

Goodale, Elaine
With careless joy we thread the woodland ways
And reach her broad domain.
Thro' sense of strength and beauty, free as air.
We feel our savage kin,—
And thus alone with conscious meaning wear
The Indian's moccasin!

All Round the Year
Moccasin Flower

MORNING-GLORY

Jackson, Helen Hunt
Wonderous interlacement!
Holding fast to threads by green and silky rings,
With the dawn it spreads its white and purple wings;
Generous in its bloom, and sheltering while it clings,
Sturdy morning-glory.

Verses
Morning Glory

Lowell, Maria White
The morning-glory's blossoming will soon be coming round;
We see their rows of heart-shaped leaves upspringing from the ground.

The Poems of Maria Lowell
The Morning-Glory
Stanza 6

MYRTLE

Montgomery, James
Dark-green and gemm'd with flowers of snow,
With close uncrowded branches spread
Not proudly high, nor meanly low,
A graceful myrtle rear'd its head.

Poetical Works of James Montgomery
Volume II
The Myrtle

NARCISSI

Plath, Sylvia
... the terrible wind tries his breathing.
The narcissi look up like children, quickly and whitely.

The Collected Poems
Among the Narcissi

ORCHID

Taylor, Bayard
Around the pillars of the palm-tree bower
The orchids cling, in rose and purple spheres;
Shield-broad the lily floats; the aloe flower
Foredates its hundred years.

The Poetical Works of Bayard Taylor
Canopus
Stanza 11

PAINTED CUP

Bryant, William Cullen
Scarlet tufts
Are glowing in the green, like flakes of fire;
The wanderers of the prairie know them well,
And call that brilliant flower the Painted Cup.

Poems
The Painted Cup

PANSY

Browning, Elizabeth Barrett
... for summer has a close,
And pansies bloom not in the snows.

The Complete Poetical Works of Elizabeth Barrett Browning
Wisdom Unapplied
Stanza II

PAPAW

Fosdick, William
And brown is the papaw's shade-blossoming cup,
In the wood, near the sun-loving maize!

Ariel and Other Poems
The Maize

PASSION FLOWER

de Vere, Sir Aubrey
Art thou a type of beauty, or of power,
Of sweet enjoyment, or disastrous sin?
For each thy name denoteth, Passion flower!

A Song of Faith
Devout Exercises and Sonnets
The Passion Flower

PEA, SWEET

Keats, John
Here are sweet peas, on tiptoe for a flight;
With wings of gentle flush o'er delicate white,
And taper fingers catching at all things,
To bind them all about with tiny rings.

Complete Poems
I Stood Tiptoe Upon a Little Hill, l. 57–60

PIMPERNEL

Thaxter, Celia
The turf is warm beneath her feet,
Bordering the beach of stone and shell,
And thick about her path the sweet
Red blossoms of the pimpernel.

The Poems of Celia Thaxter
The Pimpernel

POPPY

Bridges, Robert
A Poppy grows upon the shore

Bursts her twin cup in summer late:
Her leaves are glaucous green and hoar,
Her petals yellow, delicate.

Poetical Works of Robert Bridges
Volume II
Book I, 9

Taylor, Bayard
And far and wide, in a scarlet tide,
The poppy's bonfire spread.

The Poetical Works of Bayard Taylor
The Poet in the East
Stanza 4

PRIMROSE

Disraeli, Benjamin
"I could have brought you some primroses, but I do not like to mix violets
with anything."
"They say primroses make a capital salad," said Lord St Jerome.

Lothair
Chapter XIII (p. 57)

Wordsworth, William
A primrose by a river's brim
A yellow primrose was to him,
And it was nothing more.

The Complete Poetical Works of William Wordsworth
Peter Bell
Part I, Stanza 12

REED

Elizabeth Barrett
...those tall flowering-reeds which stand
In Arno, like a sheaf of sceptres left
But some remote dynasty of dead gods
To suck the stream for ages and get green...

The Complete Poetical Works of Elizabeth Barrett Browning
Aurora Leigh
Book VII, l. 937–40

RHODORA

Emerson, Ralph Waldo
In May, when sea-winds pierced our solitudes,
I found the fresh Rhodora in the woods,
Spreading its leafless blooms in a damp nook,
To please the desert and the sluggish brook.

Collected Poems and Translations
The Rhodora

ROSE

Bryant, William Cullen
Lovliest of lovely things are they,
On earth, that soonest pass away.
The rose that lives its little hour
Is prized beyond the sculptured flower...

Poems
A Scene on the Banks of the Hudson

Embury, Emma
The gathered rose and the stolen heart
Can charm but for a day.

The Poems of Emma C. Embury
Ballad

ROSE, WILD

Taylor, Bayard
A waft from the roadside bank
Tells where the wild rose nods.

The Poetical Works of Bayard Taylor
The Guests of Night
Stanza 2

ROSEMARY

Moore, Thomas
...the humble rosemary
Whose sweets so thanklessly are shed
To scent the desert and the dead...

The Poetical Works of Thomas Moore
Lalla Rookh
Light of the Harem (p. 259)

SAFFLOWER

Ingelow, Jean
And the saffron flower
Clear as a flame of sacrifice breaks out.

Poems
The Doom
Book II

SHAMROCK

Lover, Samuel
'll seek a four-leaf shamrock in all the fairy dells,
And if I find the charmed leaves, Oh, how I'll weave my spells!

Poems of Ireland
The Four-leaved Shamrock

SNOW-DROP

Montgomery, James
The morning star of flowers.

Poetical Works of James Montgomery
Volume II
The Snow-drop

Wordsworth, William
Nor will I then thy modest grace forget,
Chaste Snowdrops, venturous harbinger of Spring,
And pensive monitor of fleeting years!

The Complete Poetical Works of William Wordsworth
To a Snow-Drop

SPIRAEA

Goodale, Dora Read
And near the unfrequented road,
By waysides scorched with barren heat,
In clouded pink or softer white
She holds the Summer's generous light,—
Our native meadow sweet!

All Round the Year
Spiraea

SUNFLOWER

Browning, Robert
Miles and miles of golden green
Where the sunflowers blow
In a solid glow...

The Poems and Plays of Robert Browning
A Lover's Quarrel
Stanza 6

Thomson, James
But one, the lofty followers of the Sun,
Sad when he sets, shuts up her yellow leaves
Drooping all night; and, when the warm returns,
Points her enamoured bosom to his ray.

The Seasons
Summer, l. 216

SWEET BASIL

Leland, Charles G.
I pray your Highness mark this curious herb:
Touch it but lightly, stroke it softly, Sir,
And it gives forth an odour sweet and rare;
But crush it harshly and you'll make a scent
Most disagreeable.

The Music-Lesson of Confucius
Sweet Basil
Stanza 6

THORN

Wordsworth, William
There is a Thorn,—it looks so old,
In truth, you'd find it hard to say
How it could ever have been young,
It looks so old and gray.

The Complete Poetical Works of William Wordsworth
The Thorn
Stanza I

THYME

Shakespeare, William
I know a bank where the wild thyme blows.

A Midsummer-Night's Dream
Act II, Scene I, l. 249

TRILLIUM, BIRTH-ROOT

Goodale, Dora Read
Now about the rugged places
And along the ruined way,
Light and free in sudden graces
Comes the careless trend of May,—
Born of tempest, wrought in power,
Stirred by sudden hope and fear,
You may find a mystic flower
In the spring-time of the year!

All Round the Year
Trillium

TUBEROSE

Moore, Thomas
The tuberose, with her silvery light,
That in the gardens of Malay
Is call'd the Mistress of the Night,
So like a bride, scented and bright;
She comes out when the sun's away.

The Poetical Works of Thomas Moore
Lalla Rookh
Light of the Harem

TULIP

Montgomery, James
Dutch tulips from the beds
Flaunted their stately heads.

Poetical Works of James Montgomery
Volume II
The Adventure of a Star

Moore, Thomas
Like tulip-beds of different shape and dyes,
Bending beneath the invisible west-wind's sighs.

The Poetical Works of Thomas Moore
Lalla Rookh
The Veiled Prophet of Khorassan

VERBENA

Browning, Elizabeth Barrett
. . . sweet verbena which, being brushed against,
Will hold us three hours after by the smell
In spite of long walks on the windy hills.

The Complete Poetical Works of Elizabeth Barrett Browning
Aurora Leigh
Book VIII, l. 439–41

VIOLET

Bryant, William Cullen
I know where the young May violet grows,
In its lone and lowly nook.

Poems
An Indian Story
Stanza 2

Hood, Thomas
The violet is a nun;. . .

The Poetical Works of Thomas Hood
Flowers
Stanza I

WATER-LILY

Tennyson, Alfred
... the water-lily starts and slides
Upon the level in little puffs of wind,
Tho' anchor'd to the bottom...

The Complete Poetical Works of Tennyson
The Princess
IV, l. 236

WINDFLOWER

Bryant, William Cullen
The little wind-flower, whose just opened eye
Is blue as the spring heaven it gazes at...

Poems
A Winter Piece

WOODBINE

Tennyson, Alfred
And the woodbine spices are wafted abroad,
And the musk of the rose is blown.

The Complete Poetical Works of Tennyson
Maud
Part XXII, Stanza I

WORMWOOD

Thoreau, Henry David
Among the signs of autumn I perceive
The Roman wormwood (called by learned men
Ambrosia elatior, food for gods,—
For to impartial science the humblest weed
Is as immortal once as the proudest flower—)
Sprinkles its yellow dust over my shoes...

In Robert Bly
The Winged Life
Part 2
Tall Ambrosia

FORESIGHT

Meyerson, Émile

...foresight is indispensable for action. Now action for any organism of the animal kingdom is an absolute necessity. Surrounded by hostile nature it must act, it must foresee, if it wishes to live. "All life, all action," says Fouillée, "is a conscious or an unconscious divining. Divine or you will be devoured."

Identity and Reality
Chapter I (p. 22)

FUNGI

Ajello, L.
Some fungi produce a mycosis
Like blaster or histoplasmosis
But for musical sake
The one I will take
Is coccidioidomycosis.

Coccidioidomycosis
Source unknown

Holmes, Oliver Wendell
There's a thing that grows by the fainting flower,
And springs in the shade of the lady's bower;
The lily shrinks, and the rose turns pale,
When they feel its breath in the summer gale,
And the tulip curls its leaves in pride,
And the blue-eyed violet starts aside:
But the lily may flaunt, and the tulip stare,
For what does the honest toadstool care?

The Complete Poetical Works of Oliver Wendell Holmes
The Toadstool
Stanza 1

Nicholson, Norman
The toadstool towers infest the shore:
Stink-horns that propagate and spore
Wherever the wind blows.

A Local Habitation
Windscale

Peattie, Donald Culross
The fungi are the underworld of plant life, that lives clandestinely, by its unconscious wits, taking to cover in unfavorable times, rioting at another.

Flowering Earth
Chapter 18 (p. 234)

GAIA

Ehrenreich, Barbara
Some of us still get all weepy when we think about the Gaia Hypothesis, the idea that earth is a big furry goddess creature who resembles everybody's mom in that she knows what's best for us. But if you look at the historical record—Krakatoa, Mt Vesuvius, Hurricane Charley, poison ivy, and so forth down the ages—you have to ask yourself: Whose side is she on, anyway?

The Worst Years of Our Lives
The Great Syringe Tide (p. 55)

Joseph, Lawrence E.
The Gaia hypothesis is the first comprehensive scientific expression of the profoundly ancient belief that the planet Earth is a living creature.

GAIA, the Growth of an Idea
Introduction (p. 1)

GARDEN

Armstrong, Martin D.
A garden is the attempt of Man and Nature to materialize their dreams of the original Paradise. Man is its father and Nature its mother, so that all gardens which deserve the name are half-human, can appeal to us with a personality of their own.

The Atlantic Monthly
Two Italian Gardens (p. 360)
Volume CX, September 1912

Bacon, Francis
God Almighty first planted a garden.

Of Gardens
Of Gardens (p. I)

Unknown
If you would be happy for a week take a wife; if you would be happy for a month kill a pig; but if you would be happy all your life, plant a garden.

Source unknown

GENE

Beadle, George
Tatum, Ed
One gene—one enzyme.

In Francis Crick
What Mad Pursuit
The Baffling Problem (p. 33)

Benzer, Seymour
The genes are the atoms of heredity...

The Harvey Lectures
Genetic Fine Structure (p. 1)
Series 56, 1960–61

Boulding, Kenneth E.
The gene is a wonderful teacher. It is, however, a very poor learner.

The Image
Chapter 3 (p. 37)

Danforth, C.H.
One might say that the gene is to some of the biological sciences what the atom is to the physical sciences...

Science
Genetics and Anthropology (p. 216)
Volume 79, Number 2045, Friday March 9, 1934

Lote, Christopher J.
The human genome project will sooner or later identify the gene or genes responsible for religious belief. Perhaps we can contemplate (with tongue only slightly in cheek!) a brave new world in which genetic engineering

can free humanity from the scourge of religion and allow us to look forward to a bright rationalist future.

Nature
Correspondence (p. 390)
Volume 363, Number 6428, 3 June 1993

Midgley, M.
Genes cannot be selfish or unselfish, any more than atoms can be jealous, elephants abstract or biscuits teleological.

Philosophy
Gene-juggling (p. 439)
Volume 54, Number 210, 1979

Murchie, Guy
A gene is one step in the secret recipe for growing up, for living. It is a wave of the unseen wand that turns a tadpole into a frog, a caterpillar into a butterfly. It is a basic unit of heredity.

The Seven Mysteries of Life
Chapter 6 (p. 152)

Watson, James D.
We used to think our fate was in our stars. Now we know, in large measure, our fate is in our genes.

In L. Jaroff
Time
The Gene Hunt (p. 67)
Volume 133, Number 11, March 20, 1989

GENERA

Lichtenberg, Georg Christoph
In some sciences the attempt to find a general principle, an ordo, is often just as fruitless as it would be in biology to seek a general principle or a primordial particle that could have given rise to all living things. Mother Nature does not create genera and species. She creates individuals. Our nearsightedness forces us to look for similarities in order to keep everything in focus...

In E. Cramer
Chaos and Order
Chapter 1 (p. 2)

von Linne, Carl
By a botanist I mean one who understands how to observe the genera of Nature. I judge unworthy of the name of botanist the meddlesome person who is indifferent to genera.

Critica Botanica
Generic Names (p. 3)

GENETICS

Callahan, Daniel
That the emphasis has so far fallen most heavily on ridding mankind of genetic disease should not obscure the fact that the vision of genetic improvement has a lively life just below the surface, in the stirrings of a new eugenics movement.

The Tyranny of Survival
Chapter I (p. 5)

Campbell, J.H.
This new era of genetics is disclosing a remarkable new type of biological function. Some genetic structures do not adapt the organism to its environment. Instead, they have evolved to promote and direct the process of evolution. They function to enhance the capacity of the species to evolve.

In D.J. Depew and B.H. Wever (eds)
Evolution at a Crossroads: The New Biology and the New Philosophy of Science
An Organizational Interpretation of Evolution (p. 137)

Dobzhansky, Theodosius
Genetics is the first biological science which got in the position in which physics has been in for many years. One can justifiably speak about such a thing as theoretical mathematical genetics, and experimental genetics, just as in physics. There are some mathematical geniuses who work out what to an ordinary person seems a fantastic kind of theory. This fantastic kind of theory nevertheless leads to experimentally verifiable prediction, which an experimental physicist has to test the validity of. Since the times of Wright, Haldane, and Fisher, evolutionary genetics has been in a similar position.

In William B. Provine
Sewall Wright and Evolutionary Biology
Chapter 9 (p. 277)

Genetics, an important branch of biological science, has grown out of the humble peas planted by Mendel in a monastery garden.

The Rockefeller Institute Review
The Mendel Centennial
Volume 2, 1964

Huxley, Julian

In the 50 years since Mendel's Laws were so dramatically rediscovered, genetics has been transformed from a groping incertitude to a rigorous and many-sided discipline, the only branch of biology in which induction and deduction, theory and experiment, observation and comparison have come to interlock, in the same sort of way that they have for many years done in physics.

In L.C. Dunn (ed.)
Genetics in the 20th Century
Genetics, Evolution and Human Destiny (p. 591)

Koestler, Arthur

...the genetic code, written in the four-letter alphabet, 'A', 'G', 'C', 'T'. Here, then, floating in the nuclear sap, is the code which governs the skill of creating a six-foot drum major with a slight squint and dimpled cheeks, out of an egg with a diameter of a few microns.

The Act of Creation
Book 2
Chapter I (p. 417)

Lovejoy, Thomas E.

Natural species are the library from which genetic engineers can work.

Time
The Quiet Apocalypse (p. 80)
Volume 128, Number 15, 13 October 1986

Genetic engineers don't make new genes. They rearrange existing ones.

Time
The Quiet Apocalypse (p. 80)
Volume 128, Number 15, 13 October 1986

Nagle, James J.

Modern genetics is on the verge of some truly fantastic ways of "improving" the human race,...but in what direction?

Bulletin of the Atomic Scientists
Genetic Engineering
December 1971 (p. 44)

Sturtevant, A.H.

Man is one of the most unsatisfactory of all organisms for genetic study.

Science
Social Implications of the Genetics of Man (p. 405)
Volume 120, September 10, 1954

Thomas, Lewis

It is the very strangeness of nature that makes science engrossing. That ought to be at the center of science teaching. There are more than seven-times-seven types of ambiguity in science, awaiting analysis. The poetry of Wallace Stevens is crystal-clear alongside the genetic code.

Late Night Thoughts on Listening to Mahler's Ninth Symphony
Humanities and Science (p. 150)

Unknown

Why is there no life on Mars?
Because they had better genetic engineers than we do.

Source unknown

GOD

Allen, Ethan
As far as we understand nature, we are become acquainted with the character of God; for the knowledge of nature is the revelation of God.

Reason the Only Oracle of Man
Chapter I, Section II (p. 30)

Aristotle
... God and nature create nothing that has not its use.

On the Heavens
Book I, Chapter 4 271ª[30]

Blackie, John Stuart
God hath made three beautiful things,
Birds, and women, and flowers;
And he on earth who happy would be
Must look with love on all the three;
But chiefly, in bright summer hours,
He is wise who loves the flowers,
And roams the fields with me.

Musa Burschicosa
The Botanist's Song
First stanza

Buber, Martin
Nature is full of God's utterance, if one but hears it...

At the Turning Point
Third Address
Chapter IV (p. 57)

Carlyle, Thomas
[Nature] is a Volume written in celestial hieroglyphs, in a true Sacred-writing; of which even Prophets are happy that they can read here a line and there a line.

Sartor Resartus
Book III
Chapter VIII (p. 226)

...Nature, which is the Time-vesture of God, and reveals Him to the wise, hides Him from the foolish.

Sartor Resartus
Book III
Chapter VIII (p. 232)

Cowper, William
Nature is a name for an effect,
Whose cause is God.

The Poetical Works of William Cowper
The Task
The Winter Walk at Noon

Fernel, Jean
Nature embracing all things and entering into each, governs the courses and the revolutions of the sun and the moon, and of the other stars, and the succession of times, the change of the season, and the ocean's ebb and flow. Nature rules this immensity of things with an order assured and unvarying. How were it possible for Nature so to conduct and direct all this thus well but for the interposition of a divine Intelligence, which, having produced the world, preserves it?

In Sir Charles Sherrington
Man on his Nature
Chapter I (p. 21)

Graham, Aelred
Broadly speaking, all nature, since it has no claim to existence, manifests the grace of God.

Christian Thought in Action
Chapter 8 (p. 139)

Hardin, Garrett
The god who is reputed to have created fleas to keep dogs from moping over their situation must have also created fundamentalists to keep rationalists from getting flabby. Let us be duly thankful for our blessings.

> In Ashley Montagu
> *Science and Creationism*
> Introduction (p. 3)

Hobbes, Thomas
...it is impossible to make any profound inquiry into natural causes without being inclined thereby to believe there is one God eternal...

> *Leviathan*
> Part I, Chapter 11 (p. 78)

Lamarck, Jean Baptiste Pierre Antoine
...if I find that nature herself works all the wonders...that she has created organisation, life and even feeling, that she has multiplied and diversified within unknown limits the organs and faculties of the organized bodies whose existence she subserves or propagates...should I not recognise in this power of nature, that is to say in the order of existing things, the execution of the will of her Sublime Author, who was able to will that she should have this power.

> *Zoological Philosophy*
> Chapter III (p. 41)

Lawrence, D.H.
The history of the cosmos
is the history of the struggle of becoming.
When the dim flux of unformed life
struggled, convulsed back and forth upon itself,
and broke at last into light and dark
and came into existence as light,
came into existence as cold shadow
then every atom of the cosmos trembled with delight.
Behold, God is born!
He is bright light!
He is pitch dark and cold.

> *The Complete Poems of D.H. Lawrence*
> Volume II
> God is Born

Michalson, Carl
One may point to nature and say, "There *is* a God," but one cannot point to nature and say *"There* God is."

In P. Ramsey (ed.)
Faith and Ethics
Chapter 9 (p. 257)

Nicholson, Jack
When God makes a mistake, they call it nature.

In the movie
The Witches of Eastwick

Orgel, Irene
"But before Man," asked Jonah, shocked out of his wits.
"Do you mean you understood nothing at all? Didn't you exist?"
"Certainly," said God patiently. "I have told you how I exploded in the stars. Then I drifted for aeons in clouds of inchoate gas. As matter stabilized, I acquired the knowledge of valency. When matter cooled, I lay sleeping in the insentient rocks. After that I floated fecund in the unconscious seaweed upon the faces of the deep. Later I existed in the stretching paw of the tiger and the blinking eye of the owl. Each form of knowledge led to the more developed next. Organic matter led to sentience which led to consciousness which led inevitably to my divinity."
"And shall I never call you father any more? And will I never hear you call me son again?" asked Jonah.
"You may call me ," said God, agreeably, "anything you please. Would you like to discuss semantics?"

The Odd Tales of Irene Orgel
Jonah (pp. 17–18)

Pascal, Blaise
...I have a hundred times wished that if a God maintains Nature, she should testify to Him unequivocally, and that, if the signs she gives are deceptive, she should suppress them altogether.

Pensées
229

Nature has some perfections to show that she is the image of God, and some defects, to show that she is only His image.

Pensées
580

Playfair, John
The Author of nature has not given laws to the universe, which like the institutions of men, carry in themselves the elements of their own

destruction. He has not permitted, in his works, any symptom of infancy or of old age, any sign by which we may estimate either their future or their past duration.

Illustrations of the Huttonian Theory of the Earth
Section I, Part 118 (p. 119)

Temple, Frederick

The fixed laws of science can supply natural religion with numberless illustrations of the wisdom, the beneficence, the order, the beauty that characterizes the workmanship of God; while they illustrate His infinity by the marvelous complexity of natural combinations, by the variety and order of His creatures, by the exquisite finish alike bestowed on the very greatest and on the very least of His works, as if size were absolutely nothing in His sight.

Present Relations of Science to Religion (p. 13)

Thoreau, Henry David

If Nature is our mother, then God is our father.

The Writings of Henry David Thoreau
Volume I
A Week on the Concord and Merrimack Rivers
Friday (p. 492)

Whitcomb, J.
Morris, H.M.

The more we study the fascinating story of animal distribution around the earth, the more convinced we have become that this vast river of variegated life forms, moving ever outward from the Asiatic mainland, across the continents and seas, has not been a chance and haphazard phenomenon. Instead, we see the hand of God guiding and directing these creatures in ways that man, with all his ingenuity, has never been able to fathom, in order that the great commission to the postdiluvian animal kingdom might be carried out, and "that they may breed abundantly in the earth, and be fruitful, and multiply upon the earth" (Genesis 8:17).

The Genesis Flood
Chapter III (p. 86)

HEREDITY

Delage, Yves
In the field of heredity, anything is possible, and nothing is certain.

In Jean Rostand
Humanly Possible
The Evolution of Genetics (p. 100)

Hardy, Thomas
I am the family face;
Flesh perishes, I live on,
Projecting trait and trace
Through time to times anon,
And leaping from place to place
Over oblivion.

Collected Poems of Thomas Hardy
Heredity

Morgan, Thomas Hunt
That the fundamental aspect of heredity should have turned out to be so extraordinarily simple supports us in the hope that nature may, after all, be entirely approachable. Her much-advertised inscrutability has once more been found to be an illusion due to our ignorance. This is encouraging for, if the world in which we live were as complicated as some of our friends would have us believe we might well despair that biology could ever become an exact science.

The Physical Basis of Heredity
Chapter I (p. 15)

Newman, Joseph S.
Since each of us must represent
A male and female complement
Descending in unbroken line
From creatures in primordial brine,

The old hereditary genes
From worms and reptiles and sardines,
Still function in the human sperm
And still make many a man a worm.

Poems for Penguins
Heredity

HERPETOLOGISTS

Cuppy, Will

Most herpetologists have their place in the scheme of things. Because of them, we know that Butler's Garter Snake has, in most instances, only six supralabials, a state of affairs caused by the fusion of the penultimate and antepenultimate scutes. We who take our Garter Snakes so lightly may well give thought to the herpetologists counting scutes on the genus *Thamnophis* in museum basements while we are out living our lives.

How to Become Extinct
Own Your Own Snake (p. 50)

ICHTHYOLOGIST

Cuppy, Will
...it is the chief function of the ichthyologists, or fish people, to keep pointing out, day after day, the perfect fitness of fish for existence in a liquid medium. And they're right, at that. But I sometimes think that if fish were *not* well adapted for an aquatic life—if they were square, say— then it would be time to talk.

How To Become Extinct
Fish and Democracy (p. 2)

Fishback, Margaret
An ichthyologist is he,
Well versed in anthropology
To boot, so maybe he will know
Why God or nature bothered so
To give us beards and shiny noses
While fish still live on beds of roses.

I Take It Back
Fish Course

IDEA

Bernard, Claude

It is impossible to devise an experiment without a preconceived idea; devising an experiment, we said, is putting a question; we never conceive a question without an idea which invites an answer. I consider it, therefore, an absolute principle that experiments must always be devised in view of a preconceived idea, no matter if the idea be not very clear nor very well defined. As for noting the results of the experiment, which is itself only an induced observation, I posit it similarly as a principle that we must here, as always, observe without a preconceived idea.

Introduction to the Study of Experimental Medicine
Part I, Chapter I, Section vi (p. 23)

The experimental method, then, cannot give new and fruitful ideas to men who have none; it can serve only to guide the ideas of men who have them, to direct their ideas and to develop them so as to get the best possible results. The idea is a seed; the method is the earth furnishing the conditions in which it may develop, flourish, and give the best fruit according to its nature. But as only what has been sown in the ground will ever grow in it, so nothing will be developed by the experimental method except the ideas submitted to it.

Introduction to the Study of Experimental Medicine
Part I, Chapter II, Section ii (p. 34)

Bly, Robert

A great idea is a useful invention, like an eyeglass or a new fuel.

The Winged Life
Part 2 (p. 25)

Bragg, Sir Lawrence

It is not easy to be sure whether the crucial idea is really one's own or has been unconsciously assimilated in talks with others.

In J. Watson
The Double Helix
Forward by Sir Lawrence Bragg (p. viii)

Cloud, Preston

Acceptance of new ideas is usually contingent on three preconditions: (1) the world must be ready for them; (2) they must be convincingly advocated by a persuasive person or group; and (3) they must be perceived as clearly superior to (or, at least, not in serious conflict with) other widely held beliefs.

Oasis in Space: Earth History from the Beginning
Chapter 3 (p. 49)

Dewey, John

Old ideas give way slowly, for they are more than abstract forms and categories. They are habits, predispositions, deeply engrained attitudes of aversion and preference. Moreover, the conviction persists—though history shows it to be a hallucination—that all the questions that the human mind has asked are questions that can be answered in terms of the alternatives that the questions themselves present. But in fact intellectual progress usually occurs through sheer abandonment of questions together with both of the alternatives they assume—an abandonment that results from their decreasing vitality and a change of urgent interest. We do not solve them: we get over them. Old questions are solved by disappearing, evaporating, while new questions corresponding to the changed attitudes of endeavor and preference take their place. Doubtless the greatest dissolvent in contemporary thought of old questions, the greatest precipitant of new methods, new intention, new problems, is the one effected by the scientific revolution that found its climax in the *Origin of Species*.

The Influence of Darwin on Philosophy
The Influence of Darwinism on Philosophy
Section IV (p. 19)

Dobzhansky, Theodosius

Great ideas often seem simple and self-evident, but only after somebody has explained them to us. Then, how interesting they become! The act of insight is among the most exciting and pleasurable experiences a scientist can have, when he recognizes what all the time was there to be seen, and yet he did not see it.

> In Robert M. Hutchins and Mortimer J. Adler (eds)
> *The Great Ideas Today 1974*
> Advancement and Obsolescence in Science (p. 56)

Doyle, Sir Arthur Conan

One's ideas must be as broad as Nature if they are to interpret Nature.

> *The Complete Sherlock Holmes*
> A Study in Scarlet
> Part I, Chapter 5 (p. 37)

Hoefler, Don C.

Develop a honeybee mind, gathering ideas everywhere and associating them fully.

> *Electronic News*
> But You Don't Understand the Problem
> July 17, 1967

Keynes, John Maynard

The difficulty lies, not in the new ideas, but in escaping the old ones, which ramify, for those brought up as most of us have been, into every corner of our minds.

> In K.E. Drexler
> *Engines of Creation: The Coming Era of Nanotechnology* (p. 231)

Langer, Susanne K.

The limits of thought are not so much set from outside, by the fullness or poverty of experiences that meet the mind, as from within, by the power of conceptions, the wealth of formulative notions with which the mind meets experience... A new idea is a light that illuminates presences which simply had no form for us before the light fell on them.

> *Philosophy in a New Key*
> The New Key (p. 8)

Trotter, Wilfred

It is a mistake to suppose, as it is so easy to do, that science enjoins upon us the view that any given idea is true or false and there is an end of it; an idea may be neither demonstrably true nor false and yet be useful, interesting, and good exercise. Again, it is poverty rather than fertility of ideas that causes them to be used as a substitute for experiment, to be fought for with prejudice or decried with passion. When ideas are freely current they keep science fresh and living and are in no danger of ceasing to be the nimble and trusty servants of truth. We may perhaps allow ourselves to say that the body of science gets from the steady work of experiment and observation its proteins, its carbohydrates, and—sometimes too profusely—its fats, but that without its due modicum of the vitamin of ideas the whole organism is apt to become stunted and deformed, and above all to lose its resistance to the infection of orthodoxy.

British Medical Journal
Observation and Experiment and Their Use in the Medical Science (p. 132)
July 26, 1930

The mind delights in a static environment, and if there is any change to be itself the source of it. Change from without, interfering as it must with the sovereignty of the individual, seems in its very essence to be repulsive and an object of fear. A little self-examination tells us pretty easily how deeply rooted in the mind is the fear of the new, and how simple it is when fear is afoot to block the path of the new idea by unbelief and call it scepticism, and by misunderstanding and call it suspended judgment. The only way to the serene sanity which is the scientific mind—but how difficult consistently to follow—is to give to every fresh idea its one intense moment of cool but imaginative attention before venturing to mark it for rejection or suspense, as alas nine times out of ten we must do. In this traffic it is above all necessary not to be heavy-handed with ideas. It is the function of notions in science to be useful, to be interesting, to be verifiable and to acquire value from *any one* of these qualities. Scientific notions have little to gain as science from being forced into relation with that formidable abstraction "general truth". Any such relation is only too apt to discourage the getting rid of the superseded and the absorption of the new which make up the very metabolism of the mind.

British Medical Journal
The Commemoration of Great Men (p. 320)
February 20, 1932

Twain, Mark
His head was an hour-glass; it could stow an idea, but it had to do it a grain at a time, not the whole idea at once.

A Connecticut Yankee in King Arthur's Court
Chapter 28 (p. 277)

Wilde, Oscar
He played with the idea, and grew willful; tossed it into the air and transformed it; let it escape and recaptured it; made it iridescent with fancy, and winged it with paradox.

The Picture of Dorian Gray
Chapter 3 (p. 60)

IGNORANCE

Darwin, Charles

It has often and confidently been asserted, that man's origin can never be known: but ignorance more frequently begets confidence than does knowledge: it is those who know little, and not those who know so much, who so positively assert that this or that problem will never be solved by science.

The Descent of Man
Introduction (p. 253)

Franklin, Alfred

Men are enclosed in their own ignorance as in a prison with slowly receding walls. Unable to see beyond, they marvel at the vastness of their mansion without ever suspecting the existence of an infinite world outside.

In Charles Noël Martin
The Role of Perception in Science (p. 138)

IMAGINATION

Blake, William
Nature has no outline:
but Imagination has. Nature has no tune: but Imagination has!
Nature has no supernatural & dissolves: Imagination is Eternity.
The Complete Poetry and Prose of William Blake
The Ghost of Abel

The tree which moves some to tears of joy is in the Eyes of others only
a Green thing that stands in the way. Some see Nature all Ridicule &
Deformity... & Some Scarce see Nature at all. But to the Eyes of the Man
of Imagination, Nature is Imagination itself.
The Letters of William Blake
Letter to Dr Trusler
23 August 1799 (p. 34)

Born, Max
Faith, imagination and intuition are decisive factors in the progress of
science as in any other human activity.
Natural Philosophy of Cause and Chance
Appendix 1, 36 (p. 209)

Brillouin, Léon
An artist's inspiration or a scientist's theory reveal the unpredictable
power of human imagination.
Scientific Uncertainty and Information
Dedication

Harish-Chandra
I have often pondered over the roles of knowledge or experience, on the
one hand, and imagination or intuition, on the other, in the process of
discovery. I believe that there is a certain fundamental conflict between
the two, and knowledge, by advocating caution, tends to inhibit the flight

of imagination. Therefore, a certain naiveté, unburdened by conventional wisdom, can sometimes be a positive asset.

Biographical Memoirs of Fellows of the Royal Society
In R. Langlands
Harish-Chandra
Volume 31, 1985 (p. 206)

Huxley, Julian

Imagination is needed in science as much as in any other mental activity. But it must not take charge of the scientific mind. If it do, disaster may follow.

Essays in Popular Science
On the History of Science (p. 176)

Medawar, Peter

All advances of scientific understanding, at every level, begin with a speculative adventure, an imaginative preconception of *what might be true*...[This] conjecture is then exposed to criticism to find out whether or not that imagined world is anything like the real one. Scientific reasoning is, therefore, at all levels an interaction between two episodes of thought—a dialogue between two voices, the one imaginative and the other critical...

The Hope of Progress
Science and Literature (p. 16)

Pearson, Karl

Hundreds of men have allowed their imagination to solve the universe, but the men who have contributed to our real understanding of natural phenomena have been those who were unstinted in their application of criticism to the product of their imaginations.

The Grammar of Science
Introductory
Section 11 (p. 31)

Thoreau, Henry David

The imagination, give it the least license, dives deeper and soars higher than Nature goes.

The Writings of Henry David Thoreau
Volume II
Walden (p. 318)

INSECT

Clare, John
Those tiny loiterers on the barleys beard
& happy units of a numerous herd
Of playfellows the laughing summer brings
Mocking the sunshine on their glittering wings.

The Rural Muse
Insects

Eliot, George
Of what use, however, is a general certainty that an insect will not walk with his head hindmost, when what you need to know is the play of inward stimulus that sends him hither and thither in a network of possible paths?

The Writings of George Eliot
Volume 16
Daniel Deronda
Volume II
Book 3, Chapter 25 (p. 9)

Evans, Howard Ensign
The sense that insects belong to a different world than ours is shared by many people, and it is a perfectly valid feeling. After all, the search for a common ancestor of insects and ourselves would take us back more than half a billion years...In a sense insects are very much of *this* world, and *Homo sapiens* is a strange and aberrant creature of recent origin who has sought to create his own world, apart from that of nature.

The Pleasures of Entomology: Portraits of Insects and the People Who Study Them
Chapter 18 (p. 215)

Heinlein, Robert A.
In handling a stinging insect, move very slowly.

Time Enough for Love
Second Intermission (p. 365)

Kirby, William
Spence, William

...*insects*, unfortunate insects, are so far from attracting us, that we are accustomed to abhor them from our childhood. The first knowledge that we get of them is as tormentors; they are usually pointed out to us by those about us, as ugly, filthy, and noxious creatures; and the whole insect world, butterflies perhaps and some few other expected, are devoted by one universal to proscription and execration, as fit only to trodden under our feet and crushed...

An Introduction to Entomology
Introductory Letter (p. 2)

Insects, indeed, appear to have been nature's favourite productions, in which to manifest her power and skill, she has combined and concentrated almost all that is either beautiful and graceful, interesting and alluring, or curious and singular, in every other class and order of her children.

An Introduction to Entomology
Introductory Letter (p. 4)

In variegation, insects certainly exceed every other class of animated beings. Nature, in her sportive mood, when painting them, sometimes imitates the clouds of heaven; at others, the meandering course of the rivers of the earth, or the undulations of their waters: many are veined like beautiful marbles; others have the semblance of a robe of the finest net-work thrown over them; some she blazens with heraldic insignia, giving them to bear in fields sable–azure–vert–gules–argent and or fesses–bars–bends–crosses–crescents–stars, and even animals. On many, taking her rule and compasses, she draws with precision mathematical figures; points, lines, angles, triangles, squares, and circles. On others she portrays, with mystic hand, what seem like hieroglyphic symbols, or inscribes them with the characters and letters of various languages, often very correctly formed; and what is more extraordinary, she has registered in others figures which correspond with several dates of the Christian era.

An Introduction to Entomology
Introductory Letter (pp. 5–6)

Krutch, Joseph Wood

Two-legged creatures we are supposed to love as well as we love ourselves. The four-legged, also, can come to seem pretty important. But six legs are too many from the human standpoint.

The Twelve Seasons
August (p. 74)

Marquis, Don

i do not see why men

should be so proud
insects have the more
ancient lineage
according to the scientists
insects were insects
when man was only
a burbling whatisit

the lives and times of archy and mehitabel
certain maxims of archy (p. 54)

one thing that
shows that
insects are
superior to men
is the fact that
insects run their
affairs without
political campaigns
elections and so forth

the lives and times of archy and mehitabel
random thoughts by archy (p. 223)

Pallister, William
Of the INSECTS, such numbers of species exist
That their species have filled up whole volumes of books.
Over four hundred thousand are named in their list!
Every one has six legs, though they differ in looks.
The long warfare with insects gives men little ease,
For four-fifths of the whole earth's species are these!

Poems of Science
Beginnings
Animal Life (p. 140)

Teale, Edwin Way
If insects had the gift of speech, as we understand it, I am sure a main
topic of conversation would begin: 'Let me tell you about *my* molt.'

Near Horizons: The Story of an Insect Garden
Chapter 10 (p. 97)

Webb, Mary
Insects are the artists of fragrance; they have a genius for it; there seems to
be some affinity between the tenuity of their being and this most refined
of the sense-impressions.

Spring of Joy
Joy of Fragrance (p. 164)

ANT

Darwin, Charles
...the brain of an ant is one of the most marvellous atoms of matter in the world, perhaps more so than the brain of man.

The Descent of Man
Part I
Chapter II
Natural Selection (p. 281)

Hölldobler, Bert
Wilson, Edward O.
The foreign policy aim of ants can be summed up as follows: restless aggression, territorial conquest, and genocidal annihilation of neighboring colonies whenever possible. If ants had nuclear weapons, they would probably end the world in a week.

Journey to the Ants: A Story of Scientific Exploration
War and Foreign Policy (p. 59)

Proverbs 6:6
Go to the ant, thou sluggard; consider her ways, and be wise...

The Bible

Thomas, Lewis
Ants are so much like human beings as to be an embarrassment. They farm fungi, raise aphids as livestock, launch armies into war, use chemical sprays to alarm and confuse enemies, capture slaves. The family of weaver ants engage in child labor...They do everything but watch television.

The Lives of a Cell: Notes of a Biology Watcher
On Societies as Organisms (pp. 11–12)

Twain, Mark
As a thinker and planner the ant is the equal of any savage race of men; as a self-educated specialist in several arts she is the superior of any savage race of men; and in one or two high mental qualities she is above the reach of any man, savage or civilized.

What is Man?
Section 6
The Old Man
Instinct and Thought (pp. 106–7)

Science has recently discovered that the ant does not lay up anything for winter use. This will knock him out of literature, to some extent. He does not work, except when people are looking, and only then when the observer has a green, naturalistic look, and seems to be taking notes. This amounts to deception, and will injure him for the Sunday schools. He has not judgment enough to know what is good to eat from what isn't. This amounts to ignorance, and will impair the world's respect for him. He cannot stroll around a stump and find his way home again. This amounts to idiocy, and once the damaging fact is established, thoughtful people will cease to look up to him, the sentimental will cease to fondle him. His vaunted industry is but a vanity and of no effect, since he never gets home with anything he starts with. This disposes of the last remnant of his reputation and wholly destroys his main usefulness as a moral agent, since it will make the sluggard hesitate to go to him any more. It is strange beyond comprehension, that so manifest a humbug as the ant has been able to fool so many nations and keep it up so many ages without being found out.

A Tramp Abroad
Volume I
Chapter XXII (pp. 221–2)

IF WE HURRY WE MAY CATCH
DAVID ATTENBOROUGH..!

BEE

Cleveland, John
Nature's confectioner, the bee.

The Poems of John Cleveland
Fuscara, or the Bee Errant

Gay, John
The careful insect 'midst his works I view,
Now from the flow'rs exhaust the fragrant Dew;
With golden Treasures load his little Thighs,
And steer his airy Journey through the Skies;
With liquid Sweets the waxen Cells distend,
While some 'gainst Hostile Drones their Cave defend;
Each in Toil a proper Station bears,
And in the little Bulk a mighty Soul appears.

Rural Sports
Canto I, l. 83–90

Purchas the Younger, Samuel
Bees are political creatures, and destinate all their actions to one common
end; they have one common habitation, one common work, all work for
all, and one common care and love towards all their young, and that under
one Commander...

A Theatre of Political Flying-Insects (p. 16)

Shakespeare, William
...so work the honey-bees,
Creatures, that by a rule in nature, teach
The act of order to a peopled kingdom.

The Life of King Henry the Fifth
Act I, Scene ii, l. 187–9

Smythe, Daniel
The bees, those intergarden missiles,
Now make their thin propellers hum
To landing fields of flowers and thistles
More certain of their goal than some.

Nature Magazine
Small Flyers (p. 292)
Volume 50, Number 6, June–July 1957

BEETLE

Crowson, Roy A.
The beetles are at once absolutely typical of, and unique among, the Insecta, a paradox of a kind which, though familiar to any practising systematist, is a constant stumbling block to laboratory experimentalists of the modern school.

The Biology of the Coleoptera
Chapter 1 (p. 1)

Shakespeare, William
The sense of death is most in apprehension;
And the poor beetle that we tread upon,
In corporal sufferance finds a pang as great
As when a giant dies.

Measure for Measure
Act III, Scene I, l. 78–81

Wordsworth, William
The beetle, panoplied in gems and gold
A mailed angel on a battle day.

The Complete Poetical Works of William Wordsworth
Stanzas
Written In My Pocket-Copy Of Thomson's "Castle Of Indolence"

BUTTERFLY

Brower, David
"Butterfly is a stupid word," the Spaniard said; "*maripose* is so much more beautiful." I much prefer *farfalla*," the Italian countered. The woman from Paris said, "*papillon*, of course." The Japanese suggested, "I like the softness of *chocho-san*." The German bristled and demanded, "What's the matter with *schmetterling*?"

For Earth's Sake
Chapter I
Butterflies (p. 13)

Oemler, Marie Conway
There was, for instance, the common Dione Vanillae, that splendid Gulf Fritillary which haunts all the highways of the South. She's a long-wing, but she's not a Heliconian; she's a silver-spot, but she's not an Argynnis. She bears a striking family likeness to her fine relations, but she has certain structural peculiarities which differentiate her. Whose word should he take for this, and why? Wherein lay those differences? He

began, patiently, with her cylinder-shaped yellow-brown, orange-spotted caterpillar, on the purple passionflowers in our garden; he watched it change into a dark-brown chrysalis marked with a few pale spots; he saw emerge from this the red-robed lady herself, with her long fulvous fore wings, and her shorter hind wings smocked with black velvet, and under-frocked flushed with pinkish orange and spangled with silver. And yet, in spite of her long marvellous tongue—he was beginning to find out that no tool he had ever seen, and but few that God Himself makes, is so wonderful as a butterfly's tongue—she hadn't been able to tell him that about herself which he most wished to find out. *That* called for a deeper knowledge than he as yet possessed.

But he knew that other men knew. And he had to know. He meant to know. For the work gripped him as it does those marked and foreordained for its service.

Slippy Magee, Sometimes Known as Butterfly Man (pp. 72–3)

Gay, John
And what's a Butterfly? At best,
He's but a caterpillar, drest;...

The Poetical Works of John Gay
Volume II
Fable XXIV
First Volume
The Butterfly and the Snail, l. 41–2

CATERPILLAR

Isaiah 33:4
And your spoil shall be gathered like the gathering of the caterpillar.

The Bible

Unknown
A tired caterpillar went to sleep one day
In a snug little cradle of silken gray.
And he said, as he softly curled up in his nest,
"Oh, crawling was pleasant, but rest is best."

Source unknown

Walton, Izaak
And, yet, I will exercise your promised patience by saying a little of the caterpillar, or the palmer-fly or worm; that by them you may guess what a work it were, in a discourse, but to run over those very many flies, worms, and little living creatures with which the sun and summer adorn

and beautify the river-banks and meadows, both for the recreation and contemplation of us anglers; pleasures which, I think, I myself enjoy more than any other man that is not of my profession.

The Complete Angler
Part I, Chapter V, Fourth Day (p. 71)

CHIGGER

Hungerford, H.B.
The thing called a chigger
Is really no bigger
Than the smaller end of a pin.
But the bump that it raises
Just itches like blazes
And that's where the rub set in.

Attributed
In Tyler A. Woolley
Acarology
Chapter 4 (p. 41)

COCKROACH

Eliot, T.S.
She thinks that the cockroaches just need employment
To prevent them from idle and wanton destroyment.
So she's formed from that lot of disorderly louts,
A troop of well-disciplined helpful boyscouts,
With a purpose in life and a good to do—
And she's even created a Beetle Tattoo.

Old Possum's Book of Practical Cats
The Old Gumbie Cat (p. 6)

Krutch, Joseph Wood
The cockroach and the birds were both here long before we were. Both could get along very well without us, although it is perhaps significant that of the two the cockroach would miss us more.

The Twelve Seasons
November (pp. 118–19)

Marquis, Don
a good many
failures are happy
because they dont

realize it many a
cockroach believes
himself as beautiful
as a butterfly
have a heart o have
a heart and
let them dream on

> *the lives and times of archy and mehitabel*
> archygrams (p. 258)

CRICKET

Riley, James Whitcomb
But thou, O cricket, with thy roundelay,
Shalt laugh them all to scorn! So wilt thou,
pray
Trill me thy glad song o'er and o'er again:
In thy sweet prattle, since it sings the lone
Heart home again.

> *The Complete Works of James Whitcomb Riley*
> Volume III
> To the Cricket

DAMSEL-FLY

Moore, Thomas
The beautiful blue-damsel flies
That flutter'd round the jasmine stems,
Like winged flowers or flying gems...

> *The Poetical Works of Thomas Moore*
> Paradise and the Pearl, l. 409–11

DRAGONFLY

Florian, Douglas
I am the dragon,
The *demon* of skies.
Behold my bold
Enormous eyes.
I sweep
    I swoop
        I terrorize.

For lunch I munch
On flies and bees.
Mosquitoes with
My feet I seize.
I am the dragon:
Down on your knees!

Insectlopedia
The Dragonfly

FIREFLY

Beebe, William
A male firefly blazes his trail through the woods. At last he perceives a dim inconspicuous gleam, a mere spark, but it is his LANDING BEACON and he levels off and steers straight for the wingless mate, who has laboriously climbed to the top of a fern and there hung out her signal: "Come oh come, so that the race of fireflies may go on!"

High Jungle
Chapter XXI (p. 335)

Nash, Ogden
The firefly's flame
Is something for which science has no name.
I can think of nothing eerier
Than flying around with an unidentified
    red glow on a person's posterior.

Verses from 1929 On
The Firefly

FLEA

Young, Roland
And here's the happy bounding flea—
You cannot tell the he from she.
The sexes look alike you see;
But she can tell, and so can he.

Not for Children
The Flea

FLY

Doane, R.W.

A few of them [flies] were nice things to have around, to make things seem homelike...Those that were knocked into the coffee or the cream could be fished out; those that went into the soup or the hash were never missed.

Insects and Disease
Chapter V (p. 57)

Nash, Ogden

Aunt Betsy was fixing to change her will,
And would have left us out in the chill.
A *Glossina morsitans* bit Aunt Betsy
Tsk, tsk, tsetse.

Verses from 1929 On
Glossina Morsitans, or, the Tsetse

GNAT

Rudzewiz, Eugene

Gnats are gnumerous
But small.
We hardly gnotice them
At all.

In John Gardner
A Child's Beastiary
The Gnat

GRASSHOPPER

Lindsay, Vachel

The Grasshopper, the grasshopper,
I will explain to you:—
He is the Brownies' racehorse,
The fairies' Kangaroo.

Collected Poems
The Grasshopper

Lovelace, Richard

O thou that swing'st upon the waving haire
Of some well-filled Oaten Beard,
Drunke ev'ry night with a Delicious teare,
Dropt thee from Heav'n, where now th'art!

The joys of Earth and Ayre are thine intire,
That with thy feet and wings dost hop and flye;
And when thy Poppy workes, thou dost retire
To thy carv'd Acorn-bed to lye.

The Poems of Richard Lovelace
Grasse-hopper (p. 38)

KATYDID

Holmes, Oliver Wendell
Thou are a female, Katydid!
I know it by the trill
That quivers through thy piercing notes
So petulant and shrill.
I think there is a knot of you
Beneath the hollow tree,
A knot of spinster Katydids,—
Do Katydids drink tea?

The Complete Poetical Works of Oliver Wendell Holmes
To an Insect

Riley, James Whitcomb
Sometimes I keep
From going to sleep,
To hear the katydids "cheep-cheep!"
And think they say
Their prayer that way;
But *katydids* don't have to *pray*!

The Complete Works of James Whitcomb Riley
Volume VIII
The Katydids

LADY BIRD

Hurdis, James
SIR JOHN: What d'ye look at?

CECILIA: A little animal, that round my glove,
And up and down to ev'ry finger's tip,
Has travell'd merrily, and travels still,
Tho' it has wings to fly. What its name is
With learned men I know not. Simple folks

Call it the Lady-bird.

Sir Thomas More: A Tragedy
Act 1, Scene Sir Thomas More's Garden (p. 19)

MOSQUITO

Beaver, Wilfred
Mosquitoes are like little children—the moment they stop making noises you know they're getting into something.

Quote
November 10, 1968 (p. 378)

Pallister, William
The whole of Africa is our domain,
Millions of men have fought us, few remain.
We rule the lowlands of the entire earth,
The fertile lands, of far the greatest worth;
Our swarms produce their billions as we wish,
The world belongs to us, and to the fish.

Poems of Science
De Ipsa Natura
Mosquitoes (p. 219)

MOTH

Carlyle, Thomas
But see! a wandering Night-moth enters,
Allured by taper gleaming bright. . .

What passions in her small heart whirling,
Hopes boundless, adoration, dread;
At length her tiny pinions twirling,
She darts, and—puff!—the moth is dead.

In Rodger L. Tarr and Flemming McClelland (eds)
The Collected Poems of Thomas and Jane Welsh Carlyle
Tragedy of the Night-Moth
Stanza 2, 4

Wilson, Edward O.
The three-toed sloth feeds on leaves high in the canopy of the lowland forests through large portions of South and Central America. Within its fur live tiny moths, the species *Cryptoses choloepi*, found nowhere else on Earth. When a sloth descends to the forest to defecate (once a week), female moths leave the fur briefly to deposit their eggs on the fresh dung.

The emerging caterpillars build nests of silk and start to feed. Three weeks later they complete their development by turning into adult moths, and then fly up into the canopy in search of sloths. By living directly on the bodies of the sloths, the adult *Cryptoses* assure their offspring first crack at the nutrient-rich excrement and a competitive advantage over the myriad of other coprophages.

Biophilia
Bernhardsdorp (p. 9)

PRAYING MANTIS

Florian, Douglas
Upon a twig
I sit and pray
For something big
To wend my way:
A caterpillar,
Moth,
Or bee—
I swallow them
Religiously.

Insectlopedia
The Praying Mantis

TERMITE

Thomas, Lewis
When you consider the size of an individual termite, photographed standing alongside his nest, he ranks with the New Yorker and shows a better sense of organization than a resident of Los Angeles.

The Lives of a Cell: Notes of a Biology Watcher
Living Language (p. 133)

Nash, Ogden
Some primal termite knocked on wood
And tasted it, and found it good,
And that is why your Cousin May
Fell through the parlor floor today.

Verses from 1929 On
The Termite

WALKINGSTICK

Florian, Douglas
The walkingstick is thin, not thick,
And has a disappearing trick:
By looking like a twig or stalk,
It lives another day to walk.

Insectlopedia
The Walkingstick

WASP

Gay, John
Of all the plagues that heav'n hath sent,
A wasp is most impertinent!

The Poetical Works of John Gay
Volume II
Fable VIII
First Volume
The Lady and the Wasp, l. 29–30

Field, Eugene
See the wasp. He has pretty yellow stripes around his body, and a darning
needle in his tail. If you will pat the wasp upon the tail we will give you a
nice picture book.

The Complete Tribune Primer (p. 47)

WEEVIL

Florian, Douglas
We are weevils.
We are evil.
We've aggrieved
Since time Primeval.

Insectlopedia
The Weevils

INTELLIGENCE

Erskine, John

We contemplate with satisfaction the law by which in our long history one religion has driven out another, as one hypothesis supplants another in astronomy or mathematics. The faith that needs the fewest altars, the hypothesis that leaves least unexplained, survives; and the intelligence that changes most fears into opportunity is most divine. We believe this beneficent operation of intelligence was swerving not one degree from its ancient course when under the name of scientific spirits, it once more laid its influence upon religion. If the shock here seemed too violent, if the purpose of intelligence here seemed to be not revision but contradiction, it was only because religion was invited to digest an unusually large amount of intelligence all at once. Moreover, it is not certain that devout people were more shocked by Darwinism than the pious mariners were by the first boat that could tack... if intelligence begins in a pang, it proceeds to a vision.

The Moral Obligation to be Intelligent
The Moral Obligation to be Intelligent
Section V (pp. 28–30, 31)

KINGDOM

Whittaker, R.H.

There are those who consider questions in science which have no unequivocal experimentally determined answer scarcely worth discussing. Such feeling, along with conservatism, may have been responsible for the long and almost unchallenged dominance of the system of two kingdoms—plants and animals—in the broad classification of organisms. The unchallenged position of these kingdoms has ended, however; alternative systems are being widely considered.

Science
New Concepts of Kingdoms of Organisms (p. 150)
Volume 163, 10 January 1969

KNOWLEDGE

Coleridge, Mary
The fruits of the tree of Knowledge are various; he must be strong indeed who can digest all of them.

Gathered Leaves
Mary Coleridge (pp. 8–9)

Mach, Ernst
It is a peculiar property of instinctive knowledge that it is predominately of a negative nature. We cannot so well say what must happen as we can what cannot happen, since the latter alone stands in glaring contrast to the obscure mass of experience in us in which single characters are not distinguished.

The Science of Mechanics
Chapter I, Section II, 2 (p. 36)

Russell, Bertrand
Unless we can know something without knowing everything, it is obvious that we can never know something.

In John E. Leffler
Rates and Equilibria of Organic Reactions (p. v)

Severinus, Peter
...sell your lands, your house, your clothes and your jewelry; burn up your books. On the other hand, buy yourselves stout shoes, travel to the mountains, search the valleys, the deserts, the shores of the sea, and the deepest depressions of the earth; note with care the distinctions between animals, the differences of plants, the various kinds of minerals, the properties and mode of origin of everything that exists. Be not ashamed to study diligently the astronomy and terrestrial philosophy of the peasantry. Lastly, purchase coal, build furnaces, watch and operate

with the fire without wearying. In this way and no other, you will arrive at a knowledge of things and their properties.

In Allen G. Debus
The French Paracelsians
Chapter 1 (p. 8)

von Baeyer, H.C.
Field guides are instruments of the pleasure of pure knowledge.

The Sciences
Rainbows, Whirlpools, and Clouds (p. 24)
Volume 24, Number 4, July/August 1984

Wells, H.G.
The reader for whom you write
is just as intelligent as you are but
does not possess *your* store of knowledge,
he is not to be offended by a recital
in Technical language of things known to him
(e.g. telling him the position of the heart and lungs and backbone).
He is not a student preparing for an examination
& *he does not want to be*
encumbered with technical terms,
his sense of literary form & his sense of humor is probably
greater than yours.
Shakespeare, Milton, Plato, Dickens, Meredith, T.H. Huxley,
Darwin wrote for him. None of them are known to have talked
of putting in 'popular stuff' & 'treating them to pretty bits'
or alluded to matters as being 'too complicated to discuss
here'. If they were, they didn't discuss them there and *that was the end of*
it.

In Julian Huxley
Memories (p. 165)

LABORATORY

Huxley, Thomas H.
In truth, the laboratory is the fore-court of the temple of philosophy; and whoso has not offered sacrifices and undergone purification there has little chance of admission into the sanctuary.

Collected Essays
Volume VI
Hume
Part II, Chapter I (p. 61)

Thomson, W.
The laboratory of a scientific man is his place of work...The Naturalist and the botanist go to foreign lands, to study the wonders of nature, and describe and classify the results of their observations. But they must do more than merely describe, represent, and depict what they have seen. They must bring home the products of their expedition to their studies, and have recourse to the appliances of the laboratory properly so-called for their thorough and detailed examination.

Nature
Scientific Laboratories (p. 409)
Volume 31, Number 801, March 5, 1885

LARVA

Garstang, Walter
The *Amphiblastula* is notable as a larval pioneer
Who disregarded all maxims of development each year:
Half flagellate, Half amoeboid, he takes his silent plunge,
Adheres, and tucks his *front* end in, to build a baby Sponge.
Why does the *Amphiblastula* do things the wrong way round,
When hinder poles are recognized as vegetative ground?
The *Parenchymula* is worse, for he turns inside out,
His dermal cells being all inside, his gastral all without!

Larval Forms
The Amphiblastula
Stanza 1 (p. 25)

A giddy little *Gastrula*, gyrating round and round,
Was thought to show the way we got our enteron profound:
A little whirlpool in its wake maintained a tasty store,
A pocket sand to lodge it all, and left a blastopore.

Larval Forms
The Invaginate Gastrula and the Planula
Stanza 1 (p. 28)

'Tis odd that Enterozoa should with Coelenterates begin,
With differentiated cells and diplothelial skin,
For the Gastrula is clearly by a Blastula preceded,
And pelagic monothelial sires for this are sorely needed!
'Tis also true that Haeckel, when he looked around for one,
Could only *Magosphaera* find, which none else had done!
He then appealed to *Volvox*, which the serious dearth reveals,
Since both are quite incapable of taking solid meals!

Larval Forms
The Origin of Cnidoblasts and Cnidozoa
Stanza 1 (p. 30)

Leptocephalus of the Gulf Stream, the larva of the Eel,
Like a willow-leaf of clearest glass, set edgewise for a keel,
With a pair of eyes astride the stalk and tiny cleft for jaws,
That wanders for 3000 miles, two years without a pause.

Larval Forms
Leptocephalus Brevirostris
Stanza 1 (p. 66)

LAW

Camus, Albert
The laws of nature may be operative up to a certain limit, beyond which
they turn against themselves to give birth to the absurd.

The Myth of Sisyphus (p. 39)

Heinlein, Robert A.
Natural laws have no pity.

Time Enough for Love
Second Intermission (p. 369)

Huxley, Thomas H.
You have all heard it repeated, I dare say, that men of science work
by means of induction and deduction, and that by the help of these
operations, they, in a sort of sense, wring from Nature certain other things,
which are called natural laws... To hear all these large words, you would
think that the mind of a man of science must be constituted differently
from that of his fellow men; but if you will not be frightened by terms,
you will discover that you are quite wrong, and that all these terrible
apparatus are being used by yourselves every day and every hour of your
lives.

Collected Essays
Volume II
On Our Knowledge of the Causes of the Phenomena of Organic Nature
Section III (p. 364)

Mill, John Stuart
When this phraseology [Laws of Nature] was introduced, the poets
and mythologists soon took hold of it and made it subservient to their
purposes. Nature was personified: the phrase law of Nature... became a
law laid down by the goddess Nature to be obeyed by her creatures. From

the poets, this fictitious personage speedily penetrated into the closets of the philosopher...

In Ann P. Robson and John M. Robson (eds)
The Collected Works of J.S. Mill
Volume XXII
Letter to the Republican
3 January 1823 (p. 9)

Siegel, Eli
Biological laws, seen subtly, can make a girl proud.

Damned Welcome
Part II
316

Wilson, Edward O.
The laws of biology are written in the language of diversity.

BioScience
The Coming Pluralization of Biology and the Stewardship of Systematics (p. 243)
Volume 39, Number 4, April 1989

LIFE

Ardrey, Robert
As life is larger than man, so is life wiser than we are. As evolution has made us possible, so will evolution sit in final judgment. As natural selection declared us in, so natural selection should our hubris overcome us, will declare us out.

The Social Contract
The Risen Ape (p. 410)

Bernal, John Desmond
Life is a partial, continuous, progressive, multiform and conditionally interactive self-realization of the potentialities of atomic electron states...

The Origin of Life
Preface (p. xv)

Bernard, Claude
If I had to define life in a single phrase, I should clearly express my thought by throwing into relief the one characteristic which, in my opinion, sharply differentiates biological science. I should say: life is creation.

An Introduction to the Study of Experimental Medicine
Part II, Chapter II, Section I (p. 93)

It is not by struggling against cosmic conditions that the organism develops and maintains its place; on the contrary, it is by an adaptation to, an agreement with, these conditions. So, the living being does not form an exception to the great natural harmony which makes things adapt themselves to one another; it breaks no concord; it is neither in contradiction to nor struggling against general cosmic forces; far from that, it forms a member of the universal concert of things, and the life of the animal, for example, is only a fragment of the total life of the universe.

In William Maddock Bayliss
Principles of General Physiology
Preface (p. xvii)

Berrill, N.J.
Life can be thought of as water kept at the right temperature in the right atmosphere in the right light for a long enough period of time.

You and the Universe
Chapter 15 (p. 117)

Bohr, Neils
The existence of life must be considered as an elementary fact that cannot be explained, but must be taken as a starting point in biology, in a similar way as the quantum of action, which appears as an irrational element from the point of view of classical mechanical physics, taken together with the existence of elementary particles, forms the foundation of atomic physics.

Nature
Light and Life (p. 450)
Volume 131, Number 3309, April 1, 1933

Born, Max
Living matter and clarity are opposites—they run away from each other.

Letter to Albert Einstein 1927
The Born–Einstein Letters (p. 95)

Brooks, W.K.
Every reflective biologist must know that no living being is self-sufficient, or would be what it is, or would be at all, if it were not part of the natural world...Living things are real things...but their reality is in their interrelations with the rest of nature, and not in themselves.

Proceedings of the American Philosophical Society
Heredity and Variation: Logical and Biological (p. 74)
Volume 45, April 20, 1906

Crick, Francis Harry Compton
An honest man, armed with all the knowledge available to us now, could only state that in some sense, the origin of life appears to be almost a miracle, so many are the conditions which would have had to have been satisfied to get it going.

Life Itself (p. 88)

Cuvier, Georges
The development of life, the success of its forms, the precise determination of those organic types that first appeared, the simultaneous birth of certain species and their gradual extinction—the solution of these questions would perhaps enlighten us regarding the essence of the organism as much as all the experiments that we can try with living species. And man, to whom has been granted but a moment's sojourn on the earth,

would gain the glory of tracing the history of the thousands of ages which preceded his existence and of the thousands of beings that have never been his contemporaries.

In John Noble Wilford
The Riddle of the Dinosaur
Chapter 1 (p. 23)

Darwin, Erasmus

Organic Life beneath the shoreless waves
Was born, and nurs'd in Ocean's pearly caves;
First forms minute, unseen by spheric glass,
Move on the mud, or pierce the watery mass;
These, as successive generations bloom,
New powers acquire, and larger limbs assume;
Whence countless groups of vegetation spring,
And breathing realms of fin, and feet, and wing...

The Botanic Garden
Production of Life
Canto I, V, l. 295–302 (pp. 14–15)

Dawson, William Leon

I believe in love and life and beauty.

I love God and the truth and my fellow men, and the birds.

After that I love the mountains, and clouds, and flowers and shadows on the water, and pictures, and jugs (and tea-cups), and birds' eggs, and the crash of the surf, and the sighing of the wind in the pine trees, and the voices of children (babies preferred), and green things (the same as God does).

I love books—at least a good many of them. They belong to the categories preceding, for they multiply our enjoyment of everything sevenfold. I love, in spite of much cheap talk, a *Well-bound* book. Why should I submit to see Milton rigged out in fustian or Keats in jeans?

I love creative activity, the realizing of dreams, the focusing and recording of ideals, the causing of worthy things to happen.

I love the burden of the common lot, the buffet and sting of adversity, the falling down and getting up again, the clenched teeth, the still tongue—the smile over a heart which, therefore, does not break. I love the broad swept clean, and the expectation—never a certainty—that a Hand will spread it again.

I love—Oh, I love Life, *Life*, LIFE—Life as it is, Life as it may be, Life as it SHALL be.

Best of all, I love to share these things with my friends.

In Henry Chester Tracy
American Naturalists
An Ornithologist's Confession (pp. 188–9)

de Chardin, Teilhard
Man is unable to see himself entirely unrelated to mankind, neither is he able to see mankind unrelated to life, nor life unrelated to the universe.

The Phenomenon of Man
Forward (p. 34)

Doyle, Sir Arthur Conan
From a drop of water...a logician could infer the possibility of an Atlantic or a Niagara without having seen or heard of one or the other. So all life is a great chain, the nature of which is known whenever we are shown a single link of it.

The Complete Sherlock Holmes
A Study in Scarlet
Part I, Chapter 2 (p. 23)

Editorial
The fundamental distinction between the living and the non-living is that whilst it is possible to isolate the phenomena of the inorganic world, it is impossible to consider a living organism apart from its environment; it is, in fact, its reaction and adaptations to changes in its surroundings which distinguish the living from the inanimate and forms the basis of the science of biology.

Nature
Life and Death (p. 501)
Volume 122, Number 3075, October 6, 1928

Ezekiel 37:6
And I will lay sinews upon you, and will bring up flesh upon you, and cover you with skin, and put breath in you, and ye shall live.

The Bible

Garrison, W.M.
Morrison D.C.
Hamilton, J.G.
Benson, A.A.
Calvin, M.
The question of the conditions under which living matter originated on the surface of the earth is still a subject limited largely to speculation...One of the purposes of the observation reported herein is

to add another fact that might have some bearing upon this interesting question.

One of the most popular current conceptions is that life originated in an organic milieu. The problem to which we are addressed is the origin of that organic milieu in the absence of any life...

Science
Reduction of Carbon Dioxide in Aqueous Solutions by Ionizing Radiation (p. 416)
Volume 114, October 19, 1951

Haldane, J.B.S.

...Shakespeare's plays consist of words...But the arrangement of the words is even more important than the words themselves. And in the same way life is a pattern of chemical processes.

This pattern has special properties. It begets a similar pattern, as a flame does, but it regulates itself as a flame does not, except to a slight extent...So when we have said that life is a pattern of chemical processes, we have said something true and important...

But to suppose that one can describe life fully on these lines is to attempt to reduce it to mechanism, which I believe to be impossible. On the other hand, to say that life does not consist of chemical processes is to my mind as futile and untrue as to say that poetry does not consist of words.

What Is Life?
What is Life? (pp. 61–2)

Holmes, Bob

...the best minds in the world may have no problem separating the quick from the dead in ordinary experience, but they still can't agree on what life is. Living things eat, move, and excrete? So does your gas-guzzling, exhaust-belching car. Life maintains order in the face of entropy? A flame can do that. Life is the ability to replicate? Then crystals are alive but not so mules, old women and many old men.

New Scientist
Life is...? (pp. 38, 40)
Number 2138, 13 June 1998

Huxley, Julian

I turn the handle and the story starts:
Reel after reel is all astronomy,
Till life, enkindled in a niche of sky,
Leaps on the stage to play a million parts.

Essays of a Biologist
Evolution: At the Mind's Cinema (p. 2)

Krebs, H.A.
...in principle, one by one, the difficulties of explaining living systems in terms of chemistry and physics disappear. "In principle" are the operative words. In practice the difficulties remain great and seem insurmountable in the foreseeable future. But, from the point of view of the theory of knowledge, there is nevertheless a decisive difference.

Perspectives in Biology and Medicine
How the Whole becomes More Than the Sum of the Parts (p. 452)
Volume 14, Number 3, Spring 1971

Krutch, Joseph Wood
No other contrast is so tremendous as this contrast between what lives and what does not.

In R.W. Moss
Free Radical
Chapter 19 (p. 243)

Large, E.C.
There was nothing enjoyable more than a good long wrangle about plant viruses and what was meant by 'life'. But that wrangling was best left till after; until evening, when with a little alcohol to help things along, one could have a very good time, agreeing or disagreeing with each theory in turn. In the morning there was work to do.

The Advance of the Fungi
Chapter XXX (p. 416)

Leob, Jacques
Nothing indicates, however, at present that the artificial production of living matter is beyond the possibilities of science...

The Mechanistic Conception of Life
Chapter I (p. 5)

Lewis, Wyndham
Every living form is a miraculous mechanism, however, and every sanguinary, vicious or twisted need produces in Nature's workshop a series of mechanical arrangements extremely suggestive and interesting for the engineer, and almost invariably beautiful of interesting for the artist.

The Caliph's Design
The Physiognomy of Our Time (p. 77)

London, Jack
I believe that life is a mess. It is like yeast, a ferment, a thing that moves and may move for a minute, an hour, a year, or a hundred years, but that

in the end will cease to move. The big eat the little and they may continue to move, the strong eat the weak that they may retain their strength. The lucky eat the most and move the longest, that is all.

The Sea Wolf
Chapter V (p. 37)

Lorenz, Konrad
Life itself is a process of acquiring knowledge.

In P. Weiss
Hierarchically Organized Systems in Theory and Practice
Knowledge, Beliefs and Freedom (p. 231)

Mann, Thomas
What then was life? It was warmth, the warmth generated by a form-preserving insubstantiality, a fever of matter, which accompanied the process of ceaseless decay and repair of albumen molecules that were too impossibly complicated, too impossibly ingenious in structure.

The Magic Mountain
Research (p. 275)

Mather, K.F.
You would surely all agree when I assert that the mystery of the origin of life upon the face of the earth is no greater than the mystery of the origin of any single individual today upon the face of the earth.

Journal of the Scientific Laboratories of Denison University
Forty Years of Scientific Thought Concerning the Origin of Life (p. 151)
Volume 22, 1927

Mathews, Albert P.
Living things are, as it were, universes. Were it possible to magnify the human body so that the positive electrons would be as large as small shot..., a man would be about 10,000 times as tall as the distance from the earth to the sun. Were the electrons luminous, each individual would look like a nebula or collection of an immense number of suns, all of which would be in rapid orbital motion. There would be constellations, which we call molecules, and the atoms would be solar systems... We are in very truth minute universes, composed of quadrillions of suns and planets.

In E.V. Cowdry (ed.)
General Cytology
Some General Aspects of the Chemistry of Life
Section III (pp. 20, 21)

Mora, P.T.
...the presence of a living unit is exactly opposite to what we would expect on the basis of pure statistical and probability considerations.

Nature
Urge and Molecular Biology (p. 215)
Volume 199, Number 4890, July 20, 1963

Muggeridge, Malcolm
Nor, as far as I am concerned, is there any recompense in the so-called achievements of science. It is true that in my lifetime more progress has been made in unravelling the composition and the mechanism of the material universe than previously in the whole of recorded history. This does not at all excite my mind, or even my curiosity. The atom has been split; the universe has been discovered, and will soon be explored. Neither achievement has any bearing on what alone interests me—which is why life exists, and what is the significance, if any, of my minute and so transitory part in it.

In Mark Booth (ed.)
What I Believe
Malcom Muggeridge (pp. 63–4)

Muller, H.J.
To many an unsophisticated human being, the universe of stars seems only a fancy backdrop, provided for embellishing his own and his fellow creatures' performances. On the other hand, from the converse position, that of the universe of stars, not only all human beings but the totality of life is merely a fancy kind of rust, afflicting the surfaces of certain lukewarm minor planets. However, even when we admit our own littleness and the egotistical complexion of our interest in this rust, we remain confronted with the question: What is it that causes the rust to be so very fancy?

Science
Life (p. 1)
Volume 121, 7 January, 1955

Needham, James G.
...to the scientific mind the living and the non-living form one continuous series of systems of differing degrees of complexity..., while to the philosophic mind the whole universe, itself perhaps an organism, is composed of a vast number of interlacing organisms of all sizes.

Quarterly Review of Biology
Developments in Philosophy of Biology (p. 79)
Volume III, Number 1, March 1928

Peattie, Donald Culross

Whatever life is (and nobody can define it) it is something forever changing shape, fleeting, escaping us into death. Life is indeed the only thing that can die, and it begins to die as soon as it is born, and never ceases dying. Each of us is constantly experiencing cellular death. For the renewal of our tissues means a corresponding death of them, so that death and rebirth become biologically, right and left hand of the same thing. All growing is at the same time a dying away from that which lived yesterday.

The Road of a Naturalist
Chapter 12 (pp. 149–50)

Ponnamperuma, Cyril

Physicists might eventually be able to come up with a grand unification theory that encompasses not just subatomic particles and the basic elements, but the code of life as well. Who knows? Life elsewhere in the universe may even be five feet tall and standing on two legs.

In Pamela Weintraub (ed.)
The Omni Interviews
Seeds of Life (p. 3)

Sherrington, Sir Charles

A grey rock, said Ruskin, is a good sitter. That is one type of behaviour. A darting dragon-fly is another type of behaviour. We call the one alive, the other not. But both are fundamentally balances of give and take of motion with their surround. To make "life" a distinction between them is at root to treat them both artificially.

Man on His Nature
Chapter III (p. 88)

The microscope reveals that plants and animals are literally common-wealths of individually living units... Thus the corporeal house of life is built of living stones. In that house each stone is a self-centred microcosm, individually born, breathing for itself, feeding itself, consuming its own substance in its living, renewing its substance to meet that consumption, harmonizing with its own inner life some special function for the benefit of the whole, and destined ultimately for an individual death.

In T.B. Strong (ed.)
Lectures on the Method of Science
Chapter III (p. 67)

Singer, C.

We are always looking for metaphors in which to express our ideas of life, for our language is inadequate for all its complexities. Life is a labyrinth... Life is a machine... Life is a laboratory... It is but a metaphor. When we speak of ultimate things we can, maybe, speak only in metaphors. Life is a dance, a very elaborate and complex dance...

A Short History of Scientific Ideas to 1900
Chapter IX, Section 6 (p. 498)

Sinnott, E.W.

Life can be studied fruitfully in its highest as well as its lowest manifestations. The biochemist can tell us much about protoplasmic organisation, but so can the artist. Life is the business of the poet as well as of the physiologist.

Cell and Psyche
Chapter III (p. 107)

Stockbridge, Frank B.

Life is a chemical reaction; death is the cessation of that reaction; living matter, from the microscopic yeast spore to humanity itself, is merely the result of certain accidental groupings of otherwise inert matter, and *life can actually be created by repeating in the laboratory nature's own methods and processes!*

Cosmopolitan
Creating Life in the Laboratory
Volume 52, 1912 (pp. 774–81)

Szent-Györgyi, Albert

Every biologist has at some time asked 'What is life?' and none has ever given a satisfactory answer. Science is built on the premise that Nature answers intelligent questions intelligently; so if no answer exists, there must be something wrong with the question.

The Living State (p. 1)

Unknown

There was a young man of Cadiz
Who inferred that life is what it is,
For he early had learnt,
If it were what it weren't,
It could not be that which it is.

Source unknown

van Bergeijk, W.A.

Life is the necessary and sufficient condition for macromolecular systems, but macromolecules, though necessary, are not sufficient for life.

In George Gaylord Simpson
Biology and Man (p. 32)

Whitehead, Alfred North
... life is an offensive, directed against the repetitious mechanism of the Universe.

Adventures of Ideas
Chapter V (p. 102)

Wilde, Oscar
When we have fully discovered the scientific laws that govern life, we shall realize that the one person who has more illusions than the dreamer is the man of action. He, indeed, knows neither the origin of his deeds nor their results.

Intentions
The Critic as Artist, Part I

MAN

Bates, Marston
Man's point of view is curiously different in the forest and in the sea. In the forest he is a bottom animal, in the sea a surface animal.

The Forest and the Sea
Chapter 2 (p. 20)

Blake, William
Where man is not, nature is barren.

The Complete Poetry and Prose of William Blake
The Marriage of Heaven and Hell
Proverbs of Hell, l. 69

Chesterton, G.K.
...this is practically the claim of the egoism which thinks that self-assertion can obtain knowledge. A beetle may or may not be inferior to a man—the matter awaits demonstration; but if he were inferior to a man by ten thousand fathoms, the fact remains that there is probably a beetle view of things of which a man is entirely ignorant.

The Defendant
A Defence of Humility (p. 143)

De Voto, Bernard
Man is a noisome bacillus whom our Heavenly Father invented because he was disappointed in the monkey.

In Mark Twain
Mark Twain in Eruption
Introduction (p. xxvii)

Fiske, John

...Man does not dwell at the centre of things, but is the denizen of an obscure and tiny speck of cosmical matter quite invisible amid the innumerable throng of flaming suns that make up our galaxy.

The Destiny of Man
I (p. 15)

Fuller, R. Buckminster

A self-balancing, 28 jointed adapter-based biped; an electrochemical reduction-plant, integral with segregated storages of special energy extracts in storage batteries, for subsequent activation of thousands of hydraulic and pneumatic pumps, with motors attached; 62,000 miles of capillaries; millions of warning signals, railroad and conveyor systems; crushers and cranes (of which the arms are magnificent 23-jointed affairs with self-surfacing and lubricating systems, and a universally distributed telephone system (needing no service for 70 years if well-managed); the whole extraordinarily complex mechanism guided with complete precision from a turret in which are located telescopic and microscopic self-registering and recording range-finders, a spectroscope, *et cetera*, air-conditioning intake and exhaust and a main fuel intake...

Within the few cubic inches housing the turret mechanism, there is room also for two sound-wave and sound-direction-finder recording diaphragms, a filing and instant reference system, and an expertly devised analytical laboratory large enough not only to contain minute records of every last and continual event of up to 70 years experience or more, but to extend, by computation and abstract fabrication, this experience with relative accuracy into all corners of the observed universe. There is, also, a forecasting and tactical plotting department for the reduction of future possibilities and probabilities to general successful specific choice.

Nine Chains to the Moon
Chapter 4 (p. 18)

Gilman, Charlotte Perkins

Cried this pretentious Ape one day,
"I'm going to be a Man!
And stand upright, and hunt, and fight,
And conquer all I can."

Source unknown

Hugo, Victor
Man is not a circle with a single centre; he is an ellipse with two foci. Facts are one, ideas are the other.

Les Miserables
Saint-Denis
Book 7
Part I (p. 984)

Huxley, Thomas H.
The question of questions for mankind—the problem which underlies all others, and is more deeply interesting than any other is the ascertainment of the place which Man occupies in nature and of his relations to the universe of things.

Collected Essays
Volume VII
On the Relations of Man to the Lower Animals
Chapter I (p. 77)

Johanson, Donald
Edey, Maitland
...*Homo erectus*. Put him on the subway and people would probably take a suspicious look at him. Before *Homo erectus* was a really primitive type, *Homo habilis*; put him on a subway and people would probably move to the other end of the car.

Lucy: The Beginnings of Mankind
Prologue (p. 20)

James, William
Man, so far as natural science by itself is able to teach us, is no longer the final cause of the universe, the Heaven-descended heir of all the ages. His very existence is an accident, his story a brief and transitory episode in the life of one of the meanest of the planets.

Foundations of Belief
Pragmatism (p. 29)

Lawrence, D.H.
Then came the melting of the glaciers, and the world flood. The refugees from the drowned continents fled to the high places of America, Europe, Asia, and the Pacific Isles. And some degenerated naturally into cave men, neolithic and paleolithic creatures, and some retained their marvellous innate beauty and life-perfection, as the South Sea Islanders, and some wandered savage in Africa...

Fantasia of the Unconscious
Forward (p. xi)

Marquis, Don
insects have
their own point of view about
civilization a man
thinks he amounts
to a great deal
but to a
flea or a
mosquito a
human being is
merely something
to eat

the lives and times of archy & mehitabel
certain maxims of archy (pp. 53–4)

Monod, Jacques
...man knows at last that he is alone in the universe's unfeeling immensity, out of which he emerged only by chance.

Chance and Necessity
Chapter IX (p. 180)

Morris, Desmond
Despite our grandiose ideas and our lofty self-conceits, we are still humble animals, subject to all the basic laws of animal behaviour...We tend to suffer from a strange complacency that...there is something special about us, that we are somehow above biologic control. But we are not. Many exciting species have become extinct in the past, and we are no exception. Sooner or later we shall go, making way for something else. If it is to be later rather than sooner, then we must take a long, hard look at ourselves as biological specimens, and gain some understanding of our limitations.

The Naked Ape
Chapter 8 (p. 240)

Newman, Joseph S.
Man is born, eats, procreates, and dies...
This sequence of events alike applies
To horses, herring, crocodiles, and flies.

Poems for Penguins
Biochemistry

Pascal, Blaise
For, in fact, what is man in nature? A Nothingness in comparison with the Infinite, an All in comparison with the Nothing, a mean between nothing and everything.

Pensées
72

Man is but a reed, the most feeble thing in nature; but he is a thinking reed. The entire universe need not arm itself to crush him. A vapour, a drop of water suffices to kill him. But, if the Universe were to crush him, man would still be more noble than that which killed him, because he knows that he dies and the advantage which the universe has over him; the universe knows nothing of this.

Pensées
347

Russell, Bertrand
...Man is the product of causes which had no prevision of the end they were achieving; that his origin, his growth, his hopes and fears, his loves and his beliefs, are but the outcome of accidental collocations of atoms; that no fire, no heroism, no intensity of thought and feeling, can preserve an individual life beyond the grave; that all the labours of the ages, all the devotion, all the inspiration, all the noonday brightness of human genius, are destined to extinction in the vast death of the solar system, and that the whole temple of man's achievement must inevitably be buried beneath the debris of a universe in ruins—all these things, if not quite beyond dispute, are yet so nearly certain, that no philosophy which rejects them can hope to stand.

In Robert E. Egner and Lester E. Denonn
The Basic Writings of Bertrand Russell
A Free Man's Worship (p. 67)

Sackville-West, V.
...one might reply that man himself was but a collection of atoms, even as a house was but a collection of bricks, yet man laid claim to a soul, to a spirit, to a power of recording and or perception, which had not more to do with his restless atoms than had the house with its stationary bricks.

All Passion Spent
Part I (p. 82)

Simpson, George Gaylord
It is obvious that the great majority of humans throughout history have had grossly, even ridiculously, unrealistic concepts of the world. Man is, among many other things, the mistaken animal, the foolish animal. Other species doubtless have much more limited ideas about the world,

but what ideas they do have are much less likely to be wrong and are never foolish. White cats do not denigrate black, and dogs do not ask Baal, Jehovah, or other Semitic gods to perform miracles for them.

This View of Life: The World of an Evolutionist
Preface (p. viii)

Squire, J.C.
Men were on earth while climates slowly swung.
Fanning wide zones to heat and cold, and long
Subsidence turned great continents to sea,
And seas dried up, dried up interminably.
Age after age; enormous seas were dried
Amid wastes of land. And the last monster died.

Collected Poems
The Birds

Unknown
A man is an animal split halfway up and walks on the split end.

In Alexander Abingdon
Bigger & Better Boners (p. 68)

Man is a piece of the universe made alive.

Source unknown

MATHEMATICS

Comte, Auguste
In mathematics we find the primitive source of rationality; and to mathematics must the biologists resort for means to carry on their researches.

The Positive Philosophy of Auguste Compte
Volume I
Book 5, Chapter 1, To Mathematics (p. 321)

Crichton, M.
The mathematics of uncontrolled growth are frightening. A single cell of the bacterium *E. coli* would, under ideal circumstances, divide every twenty minutes. That is not particularly disturbing until you think about it, but the fact is that bacteria multiply geometrically: one becomes two, two becomes four, four becomes eight, and so on. In this way, it can be shown that in a single day, one cell of *E. coli* could produce a super-colony equal in size and weight to the entire planet earth.

The Andromeda Strain
Day 4—Spread (p. 247)

Feynman, Richard P.
To those who do not know Mathematics it is difficult to get across a real feeling as to the beauty, the deepest beauty of nature...If you want to learn about nature, to appreciate nature, it is necessary to understand the language that she speaks in.

The Character of Physical Law
Chapter 2 (p. 58)

Gold, Harvey J.
The result of a mathematical development should be continuously checked against one's own intuition about what constitutes reasonable biological behavior. When such a check reveals disagreement, then the following possibilities must be considered:

285

a. A mistake has been made in the formal mathematical development;

b. The starting assumptions are incorrect and/or constitute a too drastic oversimplification;

c.   One's own intuition about the biological field is inadequately developed;

d. A penetrating new principle has been discovered.

Mathematical Modeling of Biological Systems
Introduction
Section 1.7 (p. 15)

Haldane, J.B.S.

The permeation of biology by mathematics is only the beginning, but unless the history of science is an inadequate guide, it will continue, and the investigations here summarised represent the beginning of a new branch of applied mathematics.

The Causes of Evolution
Appendix (p. 215)

Johnson, Samuel

The Mathematicians are well acquainted with the Difference between pure Science, which has to do only with Ideas, and the Application of its Laws to the Use of Life, in which they are constrained to submit to the Imperfections of Matter and the Influence of Accidents.

The Rambler
Volume I
Number 14
May 5, 1750 (p. 80)

Koyré, A.

Nature responds only to questions posed in mathematical language, because nature is the domain of measure and order.

In H. Floris Cohen
The Scientific Revolution
Chapter 2 (p. 77)

Laplace, Pierre Simon

All the effects of nature are only the mathematical consequences of a small number of immutable laws.

In E.T. Bell
Men of Mathematics (p. 172)

Pearson, Karl
I believe the day must come when the biologist will—without being a mathematician—not hesitate to use mathematical analysis when he requires it.

Nature
Mathematics and Biology
Volume 63, Number 1629, January 17, 1901

Rapoprot, A.
The aim of mathematical biology is to introduce into the biological sciences not only quantitative, but also deductive, methods of research. The underlying idea has been to apply to biology the methods which mathematics has been successfully utilized in the physical sciences.

In H.G. Landau
Science
Mathematical Biology (p. 3)
Volume 114, July 27 1951

Stewart, Ian
Mathematics is to nature as Sherlock Holmes is to evidence.

Nature's Numbers
Chapter 1 (p. 2)

...mathematics is the science of patterns, and nature exploits just about every pattern that there is.

Nature's Numbers
Chapter 2 (p. 18)

Thompson, D'Arcy Wentworth
...the zoologist or morphologist has been slow, where the physiologist has long been eager, to invoke the aid of the physical or mathematical sciences; and the reasons for this difference lie deep...Even now the zoologist has scarce begun to dream of defining in mathematical language even the simplest organic forms.

On Growth and Form
Chapter I (p. 2)

Thoreau, Henry David
Mathematics should be mixed not only with physics but with ethics.

The Writings of Henry David Thoreau
Volume I
A Week on the Concord and the Merrimack Rivers (p. 387)

MATTER

Thompson, D'Arcy Wentworth
Cell and tissue, shell and bone, leaf and flower, are so many portions of matter, and it is in obedience to the laws of physics that their particles have been moved, molded and conformed. They are no exceptions to the rule that God always geometrizes. Their problems of form are in the first instance mathematical problems, their problems of growth are essentially physical problems, and the morphologist is, *ipso facto*, a student of physical science.

On Growth and Form
Chapter I (p. 10)

METAPHORS

Harré, Rom
Metaphor and simile are the characteristic tropes of scientific thought, not formal validity of argument.

Varieties of Realism
Part I
Introduction (p. 7)

METHOD

Billroth, Theodor
The method of research, however, of positing the questions and solving the questions posited, is invariably the same, whether we have before us a blooming rose, a diseased grape-vine, a shining beetle, the spleen of a leopard, a bird's feather, the intestines of a pig, the brain of a poet or a philosopher, a sick poodle, or a hysterical princess.

The Medical Sciences in the German Universities
Part II
The Descriptive Sciences (p. 53)

MICROBIOLOGY

Collard, Patrick
Microbiology, like all the sciences, is founded upon the twin pillars of craft technique and philosophical speculation.

The Development of Microbiology
Chapter 1 (p. 1)

MICROCOSOM

Forbes, A.
[A lake] is a little world within itself, a microcosm in which all the elemental forces are at work and the play of life goes on in full, but on a scale so small as to be easily grasped.

In F.E. Clements and V.E. Shelford
Bio-Ecology
Chapter I (p. 14)

MICROSCOPE

Baker, Henry
When you employ the Microscope, shake off all Prejudice, nor harbor any favourite Opinions; for, if you do, 'tis not unlikely Fancy will betray you into Error, and make you see what you wish to see.

The Microscope Made Easy
Chapter XV, Cautions in Viewing Objects (p. 62)

Dickens, Charles
"Yes, I have a pair of eyes," replied Sam, "and that's just it. If they was a pair o' patent double million magnifyin' gas microscopes of hextra power, p'raps I might be able to see through a flight o' stairs and a deal door; but bein' only eyes, you see my wision's limited."

Pickwick Papers
Chapter XXXIV (p. 484)

Holmes, Oliver Wendell
I was sitting with my microscope,
upon my parlor rug,
With a very heavy quarto and a very lively bug;
The true bug had been organized
with only two antennae,
But the humbug in the copperplate would have them twice as many.

The Complete Poetical Works of Oliver Wendell Holmes
Nux Postcoenatica
Stanza 1

Hooke, Robert
...me thinks it seems very probable, that nature has in these passages, as well as in those of Animal bodies, very many appropriated Instruments and contrivances, whereby to bring her designs and end to pass, which

'tis not improbable, but that some diligent Observer, if help'd with better *Microscopes*, may in time detect.

Micrographia
Observation XVIII (p. 116)

Hugo, Victor

Where the telescope ends, the microscope begins. Which of the two has the grander view?

Les Miserables
Saint-Denis
Book 3
Part III (p. 886)

Powers, Henry

Of all the Inventions none there is Surpasses
the Noble Florentine's Dioptrick Glasses
For what a better, fitter guift Could bee
in this World's Aged Luciosity.
To help our Blindnesse so as to devize
a paire of new & Articicial eyes
By whose augmenting power wee now see more
than all the world Has ever doun Before.

In S. Bradbury
The Microscope Past and Present
In Commendation of ye Microscope (p. v)

Wood, John George

...even to those who aspire to no scientific eminence, the microscope is more than an amusing companion, revealing many of the hidden secrets of Nature, and unveiling endless beauties which were heretofore enveloped in the impenetrable obscurity of their own minuteness...a good observer will discover with a common pocket magnifier many a secret of nature which has escaped the notice of a whole array of *dilettanti* microscopists in spite of all their expensive and accurate instruments.

Common Objects of the Microscope (pp. 2, 5–6)

MOLECULAR BIOLOGY

Chargaff, Erwin
... molecular biology [is] the practice of biochemistry without a license.

Essays on Nucleic Acids
Amphisbaena

Dobzhansky, Theodosius
Molecular biology is Cartesian in its inspiration.

The Biology of Ultimate Concern
Chapter 2 (p. 20)

Kornberg, Arthur
Molecular biology falters when it ignores the chemistry of the DNA blueprint—the enzymes and proteins, and their products—the integrated machinery and framework of the cell.

Biochemistry
The Two Cultures: Chemistry and Biology (p. 6890)
Volume 26, Number 22, November 3, 1987

Luria, Salvador
Molecular biology deals with questions of molecular structure, and therefore is biochemistry; but it is not the classical biochemistry that emerged earlier in the twentieth century out of the concerns of medical, agricultural, and industrial researchers. Molecular biology is genetics because it deals with genes, their functions, and their products; but, in contrast with classical genetics, it has dealt mainly with organisms such as bacteria and viruses rather than peas, maize or fruit flies, whose study had established the classical rules of genetics.

A Slot Machine, a Broken Test Tube
Chapter 4 (pp. 83–4)

Maddox, John
... coffee-breaks in molecular laboratories are as marked by speculation as in any other field, but the published literature gives the impression that

295

its authors are more concerned with the correctness of their observations than with their significance. Those with the good fortune to have the time to think about the data accumulated in the literature would probably reap a rich harvest of understanding. The explanation of the unreflective state of molecular biology is easily accounted for: competitiveness.

Nature
The Dark Side of Molecular Biology (p. 13)
Volume 363, 6 May 1993

Wolpert, Lewis
...the revolution in molecular biology changed the paradigm from metabolism to information.

The Unnatural Nature of Science
Chapter 5 (p. 93)

MOLLUSCS

Pallister, William
Next, the MOLLUSCS present forty thousand kinds more,
With a limited life, but adapted for it;
In the space of the tide, with the sea and the shore
And the sunshine, the Molluscs successfully fit;
Some on land, some in lakes which were seas, they exist
And though tideless for ages, the Molluscs persist.

Poems of Science
Beginnings
Animal Life (p. 139)

CLAM

Nash, Ogden
The clam, esteemed by gourmets highly,
Is said to live the life of Riley;
When you are lolling on a piazza
It's what you are as happy as a.

Verses from 1929 On
The Clam

NAUTILUS

Wood, Robert William
The Argo-naut or Nautilus,
With habits quite adventurous,
A com-bin-a-tion of a snail,
A jelly-fish and a paper sail.
The parts of him that did not jell,
Are packed securely in his shell.
It is not strange that when I sought

To find his double, I found Naught.

How to Tell the Birds from the Flowers and Other Wood-cuts
Naught. Nautilus. (p. 49)

OCTOPUS

Nash, Ogden
Tell me, O Octopus, I begs,
Is those things arms, or is they legs?
I marvel at thee, Octopus;
If I were thou, I'd call me Us.

Verses from 1929 On
The Octopus

OYSTER

Carroll, Lewis
"O Oyster", said the Carpenter,
"You've had a pleasant run!
Shall we be trotting home again?"
But answer came there none—
And this was scarcely odd, because
They'd eaten every one.

The Complete Works of Lewis Carroll
Through the Looking-Glass
Chapter IV (p. 188)

Twain, Mark
We know all about the habits of the ant, we know all about the habits of
the bee, but we know nothing at all about the habits of the oyster. It seems
almost certain that we have been choosing the wrong time for studying
the oyster.

The Tragedy of Pudd'nhead Wilson
Chapter XVI

SNAIL

Clare, John
There came the snail from his shell peeping out,
As fearful and cautious as thieves on the rout.

The Village Minstrel II (p. 32)

Shakespeare, William
... the snail, whose tender horns being hit,
Shrinks backward in his shelly cave with pain,
And there, all smother'd in shade, doth sit,
Long after fearing to creep forth again.

Venus and Adonis
1. 1033–6

MUSEUM

Belloc, Hilaire
The Dodo used to walk around,
And take the sun and air,
The sun yet warms his native ground—
The Dodo is not there!

The voice which used to squawk and squeak
Is now for ever dumb—
You may you see his bones and beak
All in the Mu-se-um.

Complete Verse
The Dodo

Edwards, R.Y.
The physical heart of a museum is its collection, in fact having a collection
is what makes a museum a museum, and most activity in most museums
is involved with the acquisition, care, understanding, and use of their
collections.

Occasional Papers of the British Columbia Provisional Museum
Research: A Museum Cornerstone (p. 1)
Volume 25, 1985

Flower, Sir William Henry
A museum is like a living organism; it requires constant and tender care;
it must grow or it will perish.

In Archie F. Key
Beyond Four Walls: The Origins and Development of Canadian Museums
Chapter 6 (p. 52)

Goode, George Brown

A finished museum is a dead museum, and a dead museum is a useless museum.

In Museums Association
Report of Proceedings
The Principles of Museum Administration (p. 78)
Newcastle, 1895

MUTATIONS

Crow, J.F.

...we could still be sure on theoretical grounds that mutants would usually be detrimental. For a mutation is a random change of a highly organized, reasonably smoothly functioning human body. A random change in the highly integrated system of chemical processes which constitute life is certain to impair—just as a random interchange of connections in a television set is not likely to improve the picture.

Bulletin of the Atomic Scientists
Genetic Effects of Radiation (pp. 19, 20)
Volume XIV, Number 1, January 14, 1958

Dobzhansky, Theodosius

...a majority of mutations, both those arising in laboratories and those stored in natural populations produce deteriorations of the viability, hereditary disease and monstrosities. Such changes it would seem, can hardly serve as evolutionary building blocks.

Genetics and the Origin of Species
Chapter III (p. 73)

...the mutation process alone, not corrected and guided by natural selection, would result in degeneration and extinction rather than improved adaptiveness.

American Scientist
On Methods of Evolutionary Biology and Anthropology (p. 385)
Volume 45, 1957

Huxley, Julian

One would expect that any interference with such a complicated piece of chemical machinery as the genetic constitution would result in damage. And, in fact, this is so: the great majority of mutant genes are harmful in their effects on the organism.

Evolution in Action
Chapter 2 (p. 39)

Muller, H.J.

It is entirely in line with the accidental nature of natural mutations that extensive tests have agreed in showing the vast majority of them detrimental to the organism in its job of surviving and reproducing, just as changes accidentally introduced into any artificial mechanism are predominantly harmful to its useful operation.

Bulletin of the Atomic Scientists
How Radiation Changes the Genetic Constitution (p. 331)
Volume XI, Number 9, November 1955

Pauling, Linus

Every species of plant and animal is determined by a pool of germ plasm that has been most carefully selected over a period of hundreds of millions of years. We can understand now why it is that mutations in these carefully selected organisms almost invariably are detrimental. The situation can be suggested by a statement by Dr J.B.S. Haldane: "My clock is not keeping perfect time. It is conceivable that it will run better if I shoot a bullet through it; but it is much more probable that it will stop altogether." Professor George Beadle, in this connection, has asked: "What is the chance that a typographical error would improve Hamlet?"

No More War!
Chapter 4 (p. 53)

MYRMECOLOGISTS

Hölldobler, Bert
Wilson, Edward O.
Like all myrmecologists... we are prone to view the Earth's surface idiosyncratically, as a network of ant colonies. We carry a global map of these relentless little insects in our heads. Everywhere we go their ubiquity and predictable natures makes us feel at home, for we have learned to read part of their language and we understand certain designs of their social organization better than anyone understands the behavior of our fellow humans.

Journey to the Ants: A Story of Scientific Exploration
The Dominance of Ants (p. 1)

NAMES

Borland, Hal
There is folk poetry in the common names; but science, devoted to order and systematic knowledge, insists on classifying and defining. The poet's buttercup is the botanist's Ranunculus. If you would walk with scientist as well as poet, learn both languages.

Beyond Your Doorstep
Chapter 15 (p. 359)

Carroll, Lewis
"What's the use of their having names," the Gnat said, "if they won't answer to them?"
"No use to *them*," said Alice; "but it's useful to the people that name them, I suppose."

The Complete Works of Lewis Carroll
Through the Looking-Glass
Chapter III (p. 173)

Ellis, Havelock
For even the most sober scientific investigator in science, the most thoroughgoing Positivist, cannot dispense with fiction; he must at least make use of categories, and they are already fictions, analogical fictions, or labels, which give us the same pleasure as children receive when they are told the 'name' of a thing.

The Dance of Life
The Art of Thinking (p. 89)

Ferris, G.F.
The proper aim is not to name species but to know them.

Stanford University Publications: Biological Studies
Volume 5
The Principles of Systematic Entomology

Gahan, A.B.

Objects without names cannot well be talked about or written about; without descriptions they cannot be identified and such knowledge as may have accumulated regarding them is sealed.

Entomological Society of Washington Proceedings
The Role of the Taxonomist in Present Day Entomology (p. 73)
Volume 25, 1923

Isidorus

If you know not the names, the knowledge of things is wasted.

In Carl von Linne
Critica Botanica
Generic Names (p. 1)

Page, Jake

To name something is, in a sense, to own it... [It] has been said that it is only by its name that anything can enter into thought and discourse. Naming, in other words, is a serious business.

Pastorale
What Is in a Name (p. 119)

Savory, T.
... words are in themselves among the most interesting objects of study, and the names of animals and plants are worthy of more consideration than Biologists are inclined to give them.

Naming the Living World
Preface (p. vii)

Twain, Mark
Names are not always what they seem.   The common Welsh name Bzjxxlwep is pronounced Jackson.

Following the Equator
Volume I
Chapter XXXVI
Pudd'nhead Wilson's New Calendar (p. 339)

von Linne, Carl
Name and plant are two ideas, which ought to be so closely united that they cannot possibly be separated: in order to secure this, the plant ought to lend a hand to the name, and the name in its turn to the plant, while the name in its turn rejoices in the sound principle on which it was given: since there is no connexion between botanist and plant, there is also no sound principle in naming it after him: and so the naming is bad.

Critica Botanica
Generic Names (p. 61)

For, even though the knowledge of the true and genuine Tree of Life, which might have delayed the coming of old age, is lost, still herbs remain and renew their flowers, and with perennial gratitude will always breathe forth the sweet memory of your names, and make them more enduring than marble, to outlive the names of kings and heroes.   For wealth disappears, the most magnificent houses fall into decay, the most numerous family at some time or other comes to an end: the greatest states and the most prosperous kingdoms can be overthrown: but the whole of Nature must be blotted out before the race of plants passes away, and he is forgotten who in Botany held up the torch.

Critica Botanica
Generic Names (p. 68)

The first step in wisdom is to know the things themselves; this notion consists in having the true idea of the object; objects are distinguished and known by their methodical classification and appropriate naming; therefore Classification and Naming will be the foundation of our Science.

In P.F. Stevens
The Development of Biological Systematics:
Antoine-Laurent de Jussieu, Nature and the Natural System
Chapter 9 (p. 201)

NATURAL HISTORY

Agassiz, Louis
...Natural History must in good time become the analysis of the thoughts of the Creator of the Universe as manifested in the animal and vegetable kingdoms, as well as in the inorganic world.

Essay on Classification
Chapter I, Section XXXII (p. 137)

Borlase, William
Natural History is the handmaid to Providence, collects into a narrow space what is distributed through the Universe, arranging and disposing the several Fossils, Vegetables and Animals, so as the mind may more readily examine and distinguish their beauties, investigate their causes, combinations, and effects, and rightly know how to apply them to the calls of private and public life.

The Natural History of Cornwall. The Air, Climate, Waters, Rivers, Lakes, Seas and Tides
To the Nobility and Gentry of the County of Cornwall (p. iv)

Carroll, Lewis
In one moment I've seen what has hitherto been
Enveloped in absolute mystery,
And without extra charge I will give you at large
A Lesson in Natural History.

The Hunting of the Snark
Fit the Fifth (p. 54)

Huxley, Thomas H.
To a person uninstructed in natural history, his country or sea-side stroll is a walk through a gallery filled with wonderful works of art, nine-tenths of which have their faces turned to the wall. Teach him something of natural history, and you will place in his hands a catalogue of those which are worth turning around.

Lay Sermons, Addresses, and Reviews
On the Educational Value of the Natural History Sciences (p. 91)

Smellie, William

Natural History is the most extensive, and perhaps the most instructive and entertaining of all the sciences. It is the chief source from which human knowledge is derived. To recommend the study of it from motives of utility, were to affront the understanding of mankind. Its importance, accordingly, in the arts of life, and in storing the mind with just ideas of external objects, as well as of their relations to the human race, was early perceived by all nations in their progress from rudeness to refinement.

In Buffon, Comte de Georges, Louis Leclerc
Natural History, General and Particular
Volume I
Preface by the Translator (p. ix)

NATURAL SELECTION

Bateson, William
Natural Selection is stern, but she has her tolerant moods.

In A.C. Seward
Darwin and Modern Science
Heredity and Variation in Modern Lights (p. 100)

Crick, Francis Harry Compton
Once we have become adjusted to the idea that we are here because we have evolved from simple chemical compounds by a process of natural selection, it is remarkable how many of the problems of the modern world take on a completely new light.

Molecules and Men
The Prospect Before Us (p. 93)

Darwin, Charles
We can no longer argue that, for instance, the beautiful hinge of a bivalve must have been made by an intelligent being, like the hinge of a door by man. There seems to be no more design in the variability of organic beings, and in the action of natural selection, than in the course which the wind blows.

In Francis Darwin (ed.)
Life and Letters of Charles Darwin
Volume I (p. 278)

It may metaphorically be said that natural selection is daily and hourly scrutinising, throughout the world, the slightest variations; rejecting those that are bad, preserving and adding up all that is good; silently and insensibly working, *whenever and wherever opportunity offers*, at the improvement of each organic being in relation to its organic and inorganic condition of life. We see nothing of these slow changes in progress, until the hand of time has marked the long lapse of ages...

The Origin of Species
Chapter IV (p. 42)

Slow though the process of selection may be, if feeble man can do much by artificial selection, I can see no limit to the amount of change, to the beauty and complexity of the coadaptations between all organic beings, one with another and with their physical conditions of life, which may have been effected in the long course of time through nature's power of selection...

> *The Origin of Species*
> Chapter IV
> Circumstances Favourable for the Production of
> New Forms through Natural Selection (p. 52)

...extinction and natural selection go hand in hand.

> *The Origin of Species*
> Chapter VI
> On the Absence or Rarity of Transitional Varieties (p. 80)

As natural selection acts solely by accumulating slight, successive, favourable variations, it can produce no great or sudden modification; it can act only by very short and slow steps. Hence the canon of "*Natura non facit saltum*" [Nature does not make jumps], which every fresh addition to our knowledge tends to conform, is on this theory simply intelligible.

> *The Origin of Species*
> Chapter XV (p. 235)

[Evolution by natural selection] absolutely depends on what we in our ignorance call spontaneous or accidental variability. Let an architect be compelled to build an edifice with uncut stones, fallen from a precipice. The shape of each fragment may be called accidental. Yet the shape of each has been determined...by events and circumstances, all of which depend on natural laws; but there is no relation between these laws and the purpose for which each fragment is used by the builder. In the same manner the variations of each creature are determined by fixed and immutable laws; but these bear no relation to the living structure which is slowly built up through the power of selection.

> *The Variation of Animals and Plants Under Domestication*
> Chapter XXI (p. 236)

Dawkins, Richard

All appearances to the contrary, the only watchmaker in nature is the blind forces of physics, albeit deployed in a very special way. A true watchmaker has foresight: he designs his cogs and springs, and plans their interconnections, with a future purpose in his mind's eye. Natural selection, the blind, unconscious, automatic process which Darwin discovered, and which we now know is the explanation for the existence and apparently purposeful form of all life, has no purpose in mind. It has no mind and no mind's eye. It does not plan for the future. It has no vision, no foresight, no sight at all. If it can be said to play the role of the watchmaker in nature, it is the *blind* watchmaker.

The Blind Watchmaker
Chapter 1 (p. 5)

Fisher, R.A.

Natural Selection is not Evolution. Yet, ever since the two words have been in common use, the theory of Natural Selection has been employed as a convenient abbreviation for the Theory of Evolution by means of Natural Selection...This has had the unfortunate consequences that the theory of Natural Selection itself has scarcely ever, if ever, received separate consideration.

The Genetical Theory of Natural Selection
Preface (p. vii)

Gould, Stephen Jay

The theory of natural selection would never have replaced the doctrine of divine creation if evident, admirable design pervaded all organisms. Charles Darwin understood this, and he focused on features that would be out of place in a world constructed by perfect wisdom...Darwin even wrote an entire book on orchids to argue that the structures evolved to ensure fertilization by insects are jerry-built of available parts used by ancestors for other purposes. Orchids are Rube Goldberg machines; a perfect engineer would certainly have come up with something better. This principle remains true today. The best illustrations of adaptation by evolution are the ones that strike our intuition as peculiar or bizarre.

Ever Since Darwin
Chapter 10 (p. 91)

Himmelfarb, Gertrude

Natural selection may have succeeded by default, simply because no other explanation has been available. Science, it is well known, abhors gaps as it abhors leaps, and for the same reason. The uniformity of nature and the continuum of scientific theory are both threatened by them; science's mode of knowing, its very existence, is put in jeopardy. Scientists cannot long—and a century of research is a long time as the history of modern science goes—live with the unknown, particularly when the unknown resides in the heart of their subject.

Darwin and the Darwinian Revolution
Chapter XX (pp. 366-7)

Monod, Jacques

Drawn out of the realm of pure chance, the accident enters into that of necessity, of the most implacable certainties. For natural selection operates at the macroscopic level, the level of organisms... In effect natural selection operates *upon* the products of chance and can feed nowhere else; but it operates in a domain of very demanding conditions, and from this domain chance is barred. It is not to chance but to these conditions that evolution owes its generally progressive course, its successive conquests, and the impression it gives of a smooth and steady unfolding.

Chance and Necessity
Chapter VII (pp. 118–19)

Waddington, C.H.
The meaning of natural selection can be epigrammatically summarized as 'the survival of the fittest'. Here 'survival' does not, of course, mean the bodily advance of a single individual outliving Methuselah. It implies, in its present-day interpretation, perpetuation as a source for future generations. That individual 'survives' best which leaves the most offspring. Again, to speak of an animal as 'fittest' does not necessarily imply that it is stronger or most healthy, or would win a beauty competition. Essentially it denotes nothing more than leaving most offspring. The general principle of natural selection, in fact, merely amounts to the statement that the individual which leaves most offspring are those which leave most offspring. It is a tautology.

The Strategy of the Genes
Chapter 3 (pp. 64–5)

Wallin, I.E.
Natural Selection, by itself, is not sufficient to determine the direction of organic evolution... Natural Selection can only deal with that which has been formed; it has no creative powers. Any directing influence that Natural Selection may have in organic evolution, must, in the nature of the process, be secondary to some other unknown factor.

Symbioticism and the Origin of Species
Introduction (p. 5)

Wright, R.
...natural selection "wants" us to behave in certain ways. But, so long as we comply, it doesn't care whether we are made happy or sad in the process, whether we get physically mangled, even whether we die. The only thing natural selection ultimately "wants" to keep in good shape is the information in our genes, and it will countenance any suffering on our part that serves this purpose.

The Moral Animal
Chapter 7 (p. 162–3)

NATURALIST

Darwin, Charles
It is well to remember that Naturalists value observations far more than reasoning...

In Francis Darwin (ed.)
The Life and Letters of Charles Darwin
Volume II
Darwin to Farrar
November 26, 1868 (p. 453)

A naturalist's life would be a happy one if he had only to observe, and never to write.

In Francis Darwin (ed.)
The Life and Letters of Charles Darwin
Volume II (p. 248)

Einstein, Albert
In every naturalist there must be a kind of religious feeling; for he cannot imagine that the connections into which he sees have been thought of by him for the first time. He rather has the feeling of a child, over whom a grown-up person rules.

Cosmic Religion
On Science (pp. 100–1)

Montagu, George
As natural history has, within the last half century, occupied the attention and pens of the ablest philosophers of the more enlightened parts of the globe, there needs no apology for the following sheets; since the days of darkness are now past, when the researches of the naturalist were considered as trivial and uninteresting.

Testacea Britannica
Introduction (p. I)

Riley, James Whitcomb
In gentlest worship has he bowed
To Nature. Rescued from the crowd
And din of town and thoroughfare,
He turns him from all worldly care
Unto the sacred fastness of
The forest, and the peace and love
That beats there prayer-like in the breeze...

The Complete Works of James Whitcomb Riley
Volume VII
The Naturalist

NATURE

Ackerman, Diane

Nature neither gives nor expects mercy.

The Moon by Whale Light
Chapter 4 (pp. 239–40)

Ackoff, Russell

Nature is not organized in the same way that universities are.

Management Science
Toward an Idealized University (p. B-127)
Volume 15, December 1970

Adams, Abby

Nature is what wins in the end.

The Gardener's Gripe Book
What is a Garden Anyway? (p. 10)

Agassiz, Louis

The eye of the Trilobite tells us that the sun shone on the old beach where he lived; for there is nothing in nature without a purpose; and when so complicated an organ was made to receive the light, there must have been light to enter it.

Geological Sketches
Chapter II (pp. 31–2)

As long as men inquire, they will find opportunities to know more upon these topics than those who have gone before them, so inexhaustibly rich is nature in the innermost diversity of her treasures of beauty, order, and intelligence.

Essay on Classification
Chapter II
Section I (p. 141)

Aldrich, Thomas Bailey
Nature, who loves to do a gentle thing even in her most savage moods, had taken one of those empty water-courses and filled it from end to end with forget-me-nots.

Queen of Sheba
IX (p. 205)

Aristotle
Nature has been defined as a 'principle of motion and change', and it is the subject of our inquiry. We must therefore see that we understand the meaning of 'motion'; for if it were unknown, the meaning of 'nature' too would be unknown.

Physics
Book III, Section I 200b

Arnold, Matthew
Know, man hath all which Nature hath, but more,
And in that *more* lie all his hopes of good.
Nature is cruel, man is sick of blood;
Nature is stubborn, man would fain adore.

The Poetical Works of Matthew Arnold
In Harmony with Nature

Atherton, Gertrude
Nature is a wicked old matchmaker.

Senator North
Book II, VII (p. 174)

Bacon, Francis
Nature is not governed except by obeying her.

De Aurmentis Scientiarum
Part II. Book 1, Aphorism 129

... for nature is only to be commanded by obeying her.

Novum Organum
First Book
Aphorism 129 (p. 135)

Bailey, Philip James
Nature means Necessity.

Festus: A Poem
Dedication

Baker, Henry
That Man is certainly the happiest, who is able to find out the greatest Number of reasonable and useful Amusements, easily attainable and within his Power: and, if so he that is delighted with the Works of Nature, and makes them his Study, must undoubtedly be happy, since every Animal, Flower, Fruit, or Insect, nay, almost every Particle of Matter, affords him an Entertainment.

The Microscope Made Easy
The Introduction (pp. xii–xiv)

Beston, Henry
Nature is a part of our humanity, and without some awareness and experience of that divine mystery, man ceases to be man.

The Outermost House
Forward (p. ix)

As well expect Nature to answer to your human values as to come into your house and sit in a chair.

The Outermost House
Orion Rises on the Dunes (p. 221)

Bloomfield, Robert
Strange to the world, he wore a bashful look,
The fields his study, nature was his book.

The Farmer's Boy
Spring, l. 31

Borland, Hal
There are some things, but not too many, toward which the countryman knows he must be properly respectful if he would avoid pain, sickness, and injury. Nature is neither punitive nor solicitous, but she has thorns and fangs as well as bowers and grassy banks.

Beyond Your Doorstep
Chapter 13 (p. 303)

Boyle, Robert
It is one thing to be able to help nature to produce things, and another thing to understand well the nature of the things produc'd.

The Sceptical Chymist
The Third Part (p. 95)

Bridgman, Helen Bartlett

Nature seems positively to enjoy playing pranks which turn all preconceived notions topsy-turvy.

Gems
How It Began (p. 5)

Bridgman, P.W.

...our conviction that nature is understandable and subject to law arose from the narrowness of our horizons, and that if we sufficiently extend our range we shall find that nature is intrinsically and in its elements neither understandable nor subject to law...

Harpers Magazine
The New Vision of Science (p. 444)
Volume 158, March 1929

Bronowski, Jacob

Nature is a network of happenings that do not unroll like a red carpet into time, but are intertwined between every part of the world; and we are among those parts. In this nexus, we cannot reach certainty because it is not there to be reached; it goes with the wrong model, and the certain answers ironically are the wrong answers. Certainty is a demand that is made by philosophers who contemplate the world from outside; and scientific knowledge is knowledge for action, not contemplation. There is no God's eye view of nature, in relativity, or in any science: only a man's eye view.

The Identity of Man
The Machinery of Nature
Section 6 (p. 38)

Browning, Robert

I trust in nature for the stable laws
Of beauty and utility.—Spring shall plant
And Autumn garner to the end of time...

The Poems and Plays of Robert Browning
A Soul's Tragedy
Act I (p. 458)

...what I call God.
And fools call Nature...

The Poems and Plays of Robert Browning
The Pope, l. 1073–4

Buffon, Comte de Georges, Louis Leclerc

Nature is that system of laws established by the Creator for regulating the existence of bodies, and the succession of beings. Nature is not a body; for this body would comprehend every thing. Either is it a being; for this being would necessarily be God. But nature may be considered as an immense living power, which animates the universe, and which, in subordination to the first and supreme Being, began to act by his command, and its action is still continued by his concurrence or consent.

Natural History, General and Particular
Volume VI
Of Nature
First View (p. 249)

Burroughs, John

Nature is not benevolent; Nature is just, gives pound for pound, measure for measure, makes no exceptions, never tempers her decrees with mercy, or winks at any infringement of her laws.

Harvest of a Quiet Eye
The Gospel of Nature
5 (p. 149)

Nature teaches more than she preaches. There are no sermons in stone. It is easier to get a spark out of a stone than a moral.

Time and Change
The Gospel of Nature (p. 247)

Nature exists to the mind not as an absolute realization, but as a condition, as something constantly becoming... It is suggestive and prospective; a body in motion, and not an object at rest.

The Atlantic Monthly
Expression (p. 572)
Volume VI, Number XXXVII, November 1860

Campbell, Thomas

There shall be love, when genial morn appears,
Like pensive Beauty smiling in her tears,
To watch the brightening roses of the sky,
And muse on Nature with a poet's eye.

The Complete Poetical Works of Thomas Campbell
The Pleasures of Hope
Pt II, l. 98–101

Carlyle, Thomas

Nature admits no lie; most men profess to be aware of this, but few in any measure lay it to heart.

Latter-Day Pamphlets
Number 5 (p. 170)

Nature, like the Sphinx, is of womanly celestial loveliness and tenderness; the face and bosom of a goddess, but ending in claws and the body of a lioness…Nature, Universe, Destiny, Existence, howsoever we name this grand unnamable Fact in the midst of which we live and struggle, is as a heavenly bride and conquest to the wise and brave, to them who can discern her behests and do them; a destroying fiend to them who cannot.

Past and Present
Chapter II (p. 7)

Carson, Rachel
The "control of nature" is a phrase conceived in arrogance, born of the Neanderthal age of biology and philosophy, when it was supposed that nature exists for the convenience of man.

Silent Spring
Chapter 17 (p. 297)

Chaucer, Geoffrey
Nature, the vicaire of the almyghty Lord…

The Complete Works of Geoffrey Chaucer
The Parliament of Fowls
l. 379

Chesterton, G.K.
The only words that ever satisfied me as describing Nature are the terms used in fairy books, "charm", "spell", "enchantment". They express the arbitrariness of the fact and its mystery.

Orthodoxy
The Ethics of Elfland (p. 94)

Chiras, Daniel D.
In nature, virtually nothing is wasted.

Lessons from Nature: Learning to Live Sustainably on the Earth
Chapter 2 (pp. 31–2)

Churchill, Charles
It can't be nature, for it is not sense.

The Poems of Charles Churchill
Volume II
The Farewell
l. 201

Close, Frank
Marten, Michael
Sutton, Christine
Even at subatomic level nature presents images of itself that reflect our own imaginings.

The Particle Explosion
Chapter 1 (p. 15)

Coleridge, Samuel T.
And what if all of animated nature
Be but organic harps diversely fram'd,
That tremble into thought, as o'er them sweeps,
Plastic and vast, one intellectual breeze,
At once the soul of each, and God of all?

The Complete Poetical Works of Samuel Taylor Coleridge
Volume I
The Eolian Harp
Stanza 4

All nature ministers to Hope.

Sonnets
Number 35

In nature there is nothing melancholy.

The Complete Poetical Works of Samuel Taylor Coleridge
Volume I
The Nightingale
Stanza I, l. 15

Collingwood, R.G.
The only condition on which there could be a history of nature is that the events of nature are actions on the part of some thinking being or beings, and that by studying these actions we could discover what were the thoughts which they expressed and think these thoughts for ourselves. This is a condition which probably no one will claim is fulfilled. Consequently the processes of nature are not historical processes and our knowledge of nature, though it may resemble history in certain superficial ways, e.g. by being chronological, is not historical knowledge.

The Idea of History
Part V, Section V (p. 302)

Commoner, Barry
Nature knows best.

The Closing Circle: Nature, Man & Technology
Chapter 2 (p. 41)

Cowper, William
Nature indeed looks prettily in rhyme.
The Poetical Works of William Cowper
Retirement, l. 576

da Vinci, Leonardo
Nature never breaks her own law.
Leonardo da Vinci's Notebooks
Book I
Life (p. 55)

Nature is constrained by the order of her own law which lives and works within her.
Leonardo da Vinci's Notebooks
Book I
Life (p. 55)

Necessity is the mistress and guide of nature. Necessity is the theme and artificer of nature, the bridle and the eternal law.
Leonardo da Vinci's Notebooks
Book I
Life (p. 55)

Whoever flatters himself that he can retain in his memory all the effects of Nature, is deceived, for our memory is not so capacious: therefore consult Nature for everything.
Treatise on Painting
#365 (p. 156)

Necessity is the theme and the inventress of nature, the curb and law and theme.
In Jean Paul Richter
The Literary Works of Leonardo da Vinci
Volume II
Philosophical Maxims, 1135 (p. 237)

In nature there is no effect without cause; once the cause is understood there is no need to test it by experience.
In Jean Paul Richter
The Literary Works of Leonardo da Vinci
Volume II
Philosophical Maxims, 1148B (p. 239)

Darwin, Charles
Nature will tell you a direct lie if she can.
In W.I.B. Beveridge
The Art of Scientific Investigation
Chapter 2 (p. 25)

What a book a devil's chaplain might write on the clumsy, wasteful, blundering, low, and horribly cruel works of nature.

Letter to J.D. Hooker
13 July 1856

Nature... cares nothing for appearances, except insofar as they are useful to any being. She can act on every internal organ, on every shade of constitutional difference, on the whole machinery of life.

The Origin of Species
Chapter IV (p. 41)

Darwin, Erasmus
In earth, sea, air, around, below, above,
Life's subtle woof in Nature's loom is wove,
Points glued to points in living line extends,
Touch'd by some goad approach the bending ends.

The Botanic Garden
Production of Life
Canto I, IV, l. 251–4

de Fontenelle, Bernard
...nature so entirely conceals from us the means by which her scenery is produced, that for a long time we were unable to discover the causes of her most simple movements.

Conversations on the Plurality of Worlds
First Evening (p. 9)

de Montaigne, Michel Eyquen
Let us a little permit Nature to take her own way; she better understands her own affairs than we.

The Essays
Experience III, XIII (p. 528)

Desaguliers, J.T.
Nature compell'd, his piercing Mind obeys,
And gladly shows him all her secret Ways;
'Gainst Mathematicks she has no Defence,
And yields t'experimental Consequence.

In H.N. Fairchild
Religious Trends in English Poetry
Volume I
The Newtonian System of the World (p. 357)

de Spinoza, Baruch

...Nature has set no end before herself, and that all final causes are nothing but human fictions.

Ethics
Part I
Appendix (p. 370)

Dickens, Charles

...nature gives to every time and season some beauties of its own, and from morning to night, as from the cradle to the grave, is but a succession of changes so gentle and easy, that we can scarcely mark their progress.

Nicholas Nickleby
Chapter XXII (p. 316)

Dickinson, G.L.

I'm not much impressed by the argument you attribute to Nature, that if we don't agree with her we shall be knocked on the head. I, for instance, happen to object strongly to her whole procedure: I don't much believe in the harmony of the final consummation...and I am sensibly aware of the horrible discomfort of the intermediate stages, the pushing, kicking, trampling of the host, and the wounded and dead left behind on the march. Of all this I venture to disapprove; then comes Nature and says, "But you ought to approve!" I ask why, and she says, "Because the procedure is mine." I still demur, and she comes down on me with a threat—"Very good, approve or no, as you like; but if you don't approve you will be eliminated!" "By all means," I say, and cling to my old opinion with the more affection that I feel myself invested with something of the glory of a martyr...In my humble opinion it's nature, not I, that cuts a poor figure!

The Meaning of Good
Good as the End of Nature (p. 46)

Diderot, Denis

All beings circulate from one to another; as a result all species...are in perpetual flux...All animals are more or less men; all minerals are more or less plants; all plants are more or less animals. Nothing is precise in nature.

Le Rêve de d'Alembert (pp. 69–71)

Dillard, Annie

Nature will try anything once. This is what the sign of the insects says. If you're dealing with organic compounds, then let them combine. If it works, if it quickens, set it clacking in the grass; there's always room for one more...

Pilgrim at Tinker Creek
Chapter 4, Section II (p. 65)

Dobzhansky, Theodosius

One may detest nature and despise science, but it becomes more and more difficult to ignore them. Science in the modern world is not an entertainment for some devotees. It is on its way to becoming everybody's business.

The Biology of Ultimate Concern
Chapter 1 (p. 9)

Dryden, John

For art may err, but nature cannot miss.

The Poetical Works of John Dryden
Tales From Chaucer
The Cock and the Fox, l. 452

Eckert, Allan W.

...in nature's book, everything has its place and its time; there exists a persistent interdependency of its creatures one upon another.

And there is never waste.

Wild Season
Epilogue (p. 244)

Einstein, Albert

Nature is not an engineer or contractor...

Quoted in Helen Dukas and Banesh Hoffman
Albert Einstein: The Human Side (p. 92)

Emerson, Ralph Waldo

By fate, not option, frugal Nature gave
One scent to hyson and to wall-flower,
One sound to pine-groves and to waterfalls,
One aspect to the desert and the lake.
It was her stern necessity...

Collected Poems and Translations
Xenophanes

...nature is no sentimentalist,—does not cosset or pamper us. We must see that the world is rough and surly, and will not mind drowning of a man or a woman, but swallows your ship like a grain of dust...The diseases, the elements, fortunes, gravity, lightning, respect nor persons.

The Conduct of Life
Fate (p. 4)

Nature is, what you may do...Nature is the tyrannous circumstance, the thick skull, the sheathed snake, the ponderous rock-like jaw; necessitated activity, violent direction...

The Conduct of Life
Fate (pp. 11–12)

The book of Nature is the book of Fate. She turns the gigantic pages, leaf after leaf,—never re-turning one.

The Conduct of Life
Fate (p. 12)

Nature is no spendthrift, but takes the shortest way to her ends.

The Conduct of Life
Fate (p. 32)

Nature forever puts a premium on reality.

The Conduct of Life
Behavior (p. 164)

Nature works very hard, and only hits the white once in a million throws.

The Conduct of Life
Considerations by the Way (p. 220)

Nature is a rag-merchant, who works up every shred and ort and end into new creations; like a good chemist, whom I found, the other day, in his laboratory, converting his old shirts into pure white sugar.

The Conduct of Life
Considerations by the Way (p. 230)

Nature is a mutable cloud which is always and never the same.

Essays
First Series
History (p. 18)

Nature is an endless combination and repetition of a very few laws. She hums the old well-known air through innumerable variations.

Essays
First Series
History (p. 20)

Nature is full of a sublime family likeness throughout her works and delights in startling us with resemblances in the most unexpected quarters.

Essays
First Series
History (p. 20)

Nature hates peeping, and our mothers speak her very sense when they say, "Children, eat your victuals, and say no more of it."

Essays
Second Series
Experience (p. 62)

Nature, as we know her, is no saint...She comes eating, drinking and sinning...

Essays
Second Series
Experience (p. 66)

Nature hates calculators; her methods are salatory and impulsive.

Essays
Second Series
Experience (p. 70)

To the intelligent, nature converts itself into a vast promise, and will not be rashly explained. Her secret is untold. Many and many an Oedipus arrives: he has the whole mystery teeming in his brain. Alas! The same sorcery has spoiled his skill; no syllable can he shape on his lips.

Essays
Second Series
Nature (p. 185)

Nature never hurries: atom by atom, little by little, she achieves her work.
Society and Solitude
Farming (pp. 134–5)

Nature works on a method of *all for each and each for all.*

Society and Solitude
Farming (p. 138)

Nature, like a cautious testator, fires up her estate so as not to bestow it all on one generation, but has a forelooking tenderness and equal regard to the next and the next, and the fourth and the fortieth age.

Society and Solitude
Farming (p. 139)

The great mother Nature will not quite tell her secret to the coach or the steamboat, but says, One to one, my dear, is my rule also, and I keep my enchantments and oracles for the religious soul coming alone, or as good as alone, in true-love.

Letter to Mrs Emerson
20 May 1871

Nature never wears a mean appearance. Neither does the wisest man extort her secret, and lose his curiosity by finding out all her perfection. Nature never became a toy to a wise spirit.

The Collected Works of Ralph Waldo Emerson
Volume I
Nature (p. 9)

Nature is the symbol of spirit.

The Collected Works of Ralph Waldo Emerson
Volume I
Nature
Language (p. 17)

The first steps in Agriculture, Astronomy, Zoology (those first steps which the farmer, the hunter, and the sailor take) teach that nature's dice are always loaded; that in her heaps and rubbish are concealed sure and useful results.

The Collected Works of Ralph Waldo Emerson
Volume I
Nature
Discipline (p. 25)

Evans, Howard Ensign

One's appreciation of nature is never more acute than when a bit of nature is injected into one's flesh.

The Pleasures of Entomology
Chapter 18 (p. 221)

Fermi, Enrico

Whatever nature has in store for mankind, unpleasant as it may be, men must accept, for ignorance is never better than knowledge.

In Laura Fermi
Atoms in the Family
Chapter 23 (p. 244)

Flammarion, Camille

Nature, O immense, fascinating, infinite Nature! Who can divine, who can hear, the sounds of thy celestial harmony! What can we include in these childish formulae of our young science? We lisp an alphabet while the eternal Bible is still closed to us. But it is thus when all reading begins, and these first words are surer than all the antique affirmations of ignorance and human vanity.

Popular Astronomy
Book II, Chapter III (p. 112)

Nature is immense in the little as in the great, or, to speak more correctly, for here there is neither little nor great.

Popular Astronomy
Book III, Chapter II (p. 239)

... it has been said that nature has implanted in our bosoms a craving after the discovery of truth, and assuredly that glorious instinct is never more irresistibly awakened than when our notice is directed to what is going on in the heavens.

Popular Astronomy
Book III, Chapter VII (p. 328)

Florio, John
Nature is the right law

His Firste Fruites
Proverbs
Chapter 19 (p. 32)

Foster, Sir Michael
Nature is ever making signs to us, she is ever whispering to us the beginnings of her secrets; the scientific man must be ever on the watch, ready at once to lay hold of Nature's hint, however small, to listen to her whisper, however low.

In J.A. Thomson
Introduction to Science
Chapter I (p. 16)

Gay, John
But he who studies nature's laws
From certain truth his maxims draws...

The Poetical Works of John Gay
Volume II
Introduction to the Fables, l. 76–7

Gillispie, C.C.
Indeed the renewals of the subjective approach to nature make a pathetic theme. Its ruins lie strewn like good intentions all along the ground traversed by science, until it survives only in strange corners like Lysenkoism and anthroposophy, where nature is socialized or moralized. Such survivals are relics of the perpetual attempt to escape the consequences of western man's most characteristic and successful campaign, which must doom to conquer. So like any thrust in the face of the inevitable, romantic natural philosophy has induced every nuance of mood from desperation to heroism. At the ugliest, it is sentimental or vulgar hostility to intellect. At the noblest, it inspired Diderot's naturalistic and moralizing science, Goethe's personification of nature, the poetry of Wordsworth, and the philosophy of Alfred North Whitehead, or of any other who would find a place in science for our qualitative and aesthetic appreciation of nature. It is the science of those who would make botany of blossoms and meteorology of sunsets.

The Edge of Objectivity
Chapter V (pp. 199–200)

Gregg, Alan
One wonders whether the rare ability to be completely attentive to, and to profit by, nature's slightest deviation from the conduct expected of her is

not the secret of the best research minds and one that explains why some men turn to most remarkably good advantage seemingly trivial accidents. Behind such attention lies an unremitting sensitivity, analogous, I suspect, to that strange experience we all have of encountering a new word two or three times within the first few weeks after we have learned it.

The Furtherance of Medical Research
Chapter III (p. 98)

Gregory, Dick

Nature is not affected by finance. If someone offered you ten thousand dollars to let them touch you on your eyeball without your blinking, you would never collect the money. At the very last moment, Nature would force you to blink your eye. Nature will protect her own.

The Shadow that Scares Me
Chapter VIII
Nature Protects Her Own (p. 175)

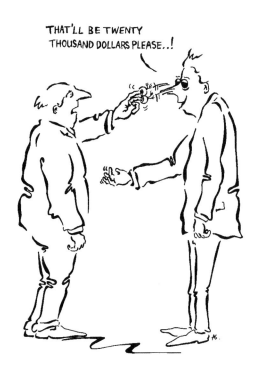

Haeckel, Ernst
The anthropomorphic notion of a deliberate architect and ruler of the world has gone forever from this field; the "eternal iron laws of nature" have taken its place.

The Riddle of the Universe
Chapter XIV (p. 261)

Hales, Stephen
...our reasonings about the wonderful and intricate operations of Nature are so full of uncertainty, that, as the wise-man truly observes, *hardly do we guess aright at the things that are upon earth, and with labour do we find the things that are before us.*

Vegetable Staticks
Chapter VII (p. 181)

Henley, W.E.
What Nature has writ with her lusty wit
  Is worded so wisely and kindly
That whoever has dipped in her manuscript
  Must up and follow her blindly.

Echoes of Life and Death
Number XXXIII

Heraclitus
Nature loves to hide.

Fragments
Fragment X (p. 4)

Horace
You may drive out nature with a pitchfork, yet she will ever hurry back.

Satires, Epistles, and Ars Poetica
Epistles
Book I, Epistle 10, l. 23

Huxley, Thomas H.
Nature is never in a hurry, and seems to have had always before her eyes the adage, "keep a thing long enough and you will find a use for it."

Collected Essays
Volume VIII
Discourses, Biological and Geological

James, William
It seems *a priori* improbable that the truth should be so nicely adjusted to our needs and powers... In the great boarding-house of nature, the cakes

and the butter and the syrup seldom come out so even and leave the plates so clean.

The Will to Believe and Other Essays in Popular Philosophy
The Will to Believe
VIII (p. 27)

Visible nature is all plasticity and indifference,—a moral multiverse... and not a moral universe. To such a harlot we owe no allegiance; with her as a whole we can establish no moral communion; and we are free in our dealing with her several parts to obey or to destroy, and to follow no law but that of prudence in coming to terms with such of her particular features as will help us to our private ends.

The Will to Believe and Other Essays in Popular Philosophy
Is Life Worth Living? (p. 43)

Janzen, Daniel

Here's what nature does for us, no matter who we are or where we live. Human animals carry around this big brain, this big device for processing input. Part of our ability to use that device depends on the complex stimuli that challenged it throughout our evolution. Nature—whatever is out there, from a single tree to a whole forest—provides a big wad of the possible information we can process. If you diminish nature, you diminish the diversity of those stimuli. When we don't get input from nature, we end up having not much sense of smell, hearing, vision. Television becomes our reality. We can survive on that, and do. But it's not nearly as complex. I'll always put my money on the person who has the broadest background—I don't care whether it's in art, music, finance, or nature. I'm not arguing against cities and all they have to offer, but life is bigger than that, more than that. When we diminish nature, we turn off a lot of things in our own heads. People should care about nature for a very practical reason—the more experience we have, the better off we are.

Over the past ten or fifteen years, I've been bothered by the fact that Americans think they're getting nature through TV—all those shows that bring the elephants and tigers right into the living rooms. This Muzak nature destroys the reality of people's experience outdoors. When they're actually in nature, it's disappointing, because the big, spectacular stimuli aren't coming as fast as they do on television. In nature, you might have to wait six hours or six days to see that bird. People who haven't been immersed in that TV background are much more affected when they visit the tropical forest.

In Winifred Gallagher
The Power of Place
Chapter 14 (pp. 207–8)

Krutch, Joseph Wood
To those who study her, Nature reveals herself as extraordinarily fertile and ingenious in devising *means*, but she has no *ends* which the human mind has been able to discover or comprehend.

The Modern Temper
Chapter II, Section III (p. 27)

Laird, John
There is beauty... in sky and cloud and sea, in lilies and in sunsets, in the glow of bracken in autumn and in the enticing greenness of a leafy spring. Nature, indeed, is infinitely beautiful, and she seems to wear her beauty as she wears colours or sound. Why then should her beauty belong to us rather than to her?

A Study in Realism
Chapter VII (p. 129)

Lamarck, Jean Baptiste Pierre Antoine
Nature has produced all the species of animals in succession, beginning with the most imperfect or simplest, and ending her work with the most perfect, so as to create a gradually increasing complexity in their organisation; these animals have spread at large throughout all the habitable regions of the globe, and every species has derived from its environment the habits that we find in it and the structural modifications which observation shows us.

Zoological Philosophy
Chapter VII (p. 126)

Lewis, C.S.
If ants had a language they would, no doubt, call their anthill an artifact and describe the brick wall in its neighbourhood as a *natural* object. Nature in fact would be for them all that was not 'ant-made'. Just so, for us, *nature* is all that is not man-made; the natural state of anything is its state when not modified by man.

Studies in Words
Nature (pp. 45–6)

Longfellow, Henry Wadsworth
No tears
Dim the sweet look that Nature wears.

The Complete Writings of Henry Wadsworth Longfellow
Volume I
Sunrise on the Hills

Marsh, George Perkins

Nature, left undisturbed, so fashions her territory as to give it almost unchanging permanence of form, outline, and proportion, except when shattered by geologic convulsions; and in these comparatively rare cases of derangement, she sets herself at once to repair the superficial damage, and to restore, as nearly as practicable, the former aspect of her dominion.

Man and Nature; or, Physical Geography as Modified by Human Action
Introductory (p. 29)

McKibben, Bill

The end of nature sours all my material pleasures. The prospect of living in a genetically engineered world sickens me. And yet it is toward such a world that our belief in endless material advancement hurries us. As long as that desire drives us, here is no way to set limits.

The End of Nature
A Path of More Resistance (p. 173)

Mill, John Stuart

Nature means the sum of all phenomena, together with the causes which produce them; including not only all that happens, but all that is capable of happening...

Three Essays on Religion
Nature (p. 5)

Morley, John

Nature, in her most dazzling aspects or stupendous parts, is but the background and theatre of the tragedy of man.

Critical Miscellanies
Byron (p. 140)

Muir, John

One is constantly reminded of the infinite lavishness and fertility of Nature—inexhaustible abundance amid what seems enormous waste. And yet when we look into any of her operations that lie within reach of our minds, we learn that no particle of her material is wasted or worn out. It is eternally flowing from use to use, beauty to yet higher beauty.

Gentle Wilderness (p. 139)

Nature is a good mother, and sees well to the clothing of her many bairns—birds with smoothly imbricated feathers, beetles with shining jackets, and bears with shaggy furs. In the tropical south, where the sun warms like a fire, they are allowed to go thinly clad; and in the snowy northland she takes care to clothe warmly. The squirrel has socks and mittens, and a tail broad enough for a blanket; the grouse is densely feathered down to the ends of his toes; and the wild sheep, besides his undergarment of fine wool, has a thick overcoat of hair that sheds off both the snow and the rain. Other provisions and adaptations in the dresses of animals, relating less to climate than to the more mechanical circumstances of life, are made with the same consummate skill that characterizes all the love-work of Nature.

> In Sally M. Miller (ed.)
> *John Muir: Life and Work*
> Part III, Chapter 5 (pp. 111–12)

Newton, Sir Isaac
For Nature is very consonant and conformable to herself.

> *Optics*
> Book III, Part 1, Question 31 (p. 531)

Oliver, Mary
Nature, the total of all of us, is the wheel that drives our world; those who ride it willingly might yet catch a glimpse of a dazzling, even a spiritual restfulness, while those who are unwilling simply to hang on, who insist that the world must be piloted by man for his own benefit, will be dragged around and around all the same, gathering dust but no joy.

> *Blue Pastures*
> A Few Words (p. 92)

Peters, Ted
Nature as we daily experience it is ambiguous, fraught with benefits and liabilities.

> *Playing God?*
> Playing God with DNA (p. 20)

Poincaré, Henri
The scientist does not study nature because it is useful; he studies it because he delights in it, and he delights in it because it is beautiful. If nature were not beautiful, it would not be worth knowing, and if nature were not worth knowing, life would not be worth living.

> *The Foundations of Science*
> Science and Method
> The Choice of Facts (p. 366)

Pope, Alexander
All nature is but art, unknown to thee.

> *Alexander Pope's Collected Poems*
> Essay on Man
> Epistle I, l. 289

Those rules of old, discover'd, not devis'd,
Are Nature still, but Nature methodized;
Nature, like liberty, is but restrain'd
By the same laws which first herself ordain'd.

Alexander Pope's Collected Poems
An Essay on Criticism
Part I, l. 88–91

Quammen, David

Nature grants no monopolies in resourcefulness. She does not even seem to hold much with the notion of portioning it out hierarchically. Gold, she decrees, is where you find it.

Natural Acts
A Better Idea (p. 3)

Reade, Winwood

When we have ascertained by means of Science, the method of Nature's operations, we shall be able to take her place and perform them for ourselves.

The Martyrdom of Man
Chapter IV
Inventions of the Future (p. 513)

Richet, Charles

Nature guards her secrets jealously: it is necessary to lay violent siege to her for a long time to discover a single one of them, however small it be.

The Natural History of a Savant
Chapter XIII (p. 149)

Saunders, W.E.

Lovers of nature feel so confidently that their hobby is an enormous asset in life that there is no feeling of hesitancy in advocating that every person should become acquainted with new species of birds, trees, insects, etc., just as often as opportunity offers. And the time to do so is always NOW!

In R.J. Rutter (ed.)
W.E. Saunders—Naturalist
Saunderisms (p. 50)

Scott, Walter

Some touch of Nature's genial glow...

The Lord of the Isles
Canto III, stanza XIV

Sears, Paul
Nature is not to be conquered save on her own terms. She is not conciliated by cleverness or industry in devising means to defeat the operation of one of her laws through the workings of another.

Deserts on the March
Chapter I (p. 3)

Seneca
It is difficult to change nature.

De Ira
II

A day will come in which zealous research over long periods of time will bring to light things that now still lie hidden. The life of a single man, even if he devotes it entirely to the heavens, is insufficient to fathom so broad a field. Knowledge will thus unfold only over the course of generations. But there will come a time when our descendants will marvel that we did not know the things that seem so simple to them. Many discoveries are reserved for future centuries, however, when we are long forgotten. Our universe would be deplorably insignificant had it not offered every generation new problems. Nature does not surrender her secrets once and for all.

Naturales Quaestiones
Book 7

Shakespeare, William
In Nature's infinite book of secrecy
A little I can read.

Anthony and Cleopatra
Act I, Scene 2, l. 9–10

One touch of nature makes the whole world kin.

Troilus and Cressida
Act III, Scene III, l. 175

Thou, nature, art my goddess; to thy laws
My services are bound.

King Lear
Act I, Scene II, l. 1–2

Spencer, Herbert
Nature's rules... have no exceptions.

Social Statics
Introduction
Lemma II (p. 39)

Stevenson, Adlai E.
Nature is neutral. Man has wrested from nature the power to make the world a desert or to make the deserts bloom.

Speech
Hartford, Connecticut
September 18, 1952

Stillingfleet, Benjamin
. . . each moss,
Each shell, each crawling insect holds a rank
Important in the plan of Him, who fram'd
This scale of beings; holds a rank, which lost
Wou'd break the chain, and leave behind a gap
Which nature's self would rue.

Miscellaneous Tracts to Natural History, Husbandry, and Physick (p. 128)

Swann, W.F.C.
There are times. . . in the growth of human thought when nature, having led man to the hope that he may understand her glories, turns for a time capricious and mockingly challenges his powers to harmonize her mysteries by revealing new treasures.

In Bernard Jaffe
Crucibles
Chapter XVI (p. 322)

Swift, Jonathan
He said, that new systems of nature were but new fashions, which would vary in every age; and even those who pretend to demonstrate them from mathematical principles, would flourish but a short period of time, and be out of vogue when that was determined.

Gulliver's Travels
Part III, Chapter VIII (pp. 118–19)

Teale, Edwin Way
Nature is shy and noncommittal in a crowd. To learn her secrets, visit her alone or with a single friend, at most.

Circle of the Seasons
May 4 (p. 85)

Tennyson, Alfred
Who trusted
God was love indeed
And love Creation's final law—
Tho' Nature, red in tooth and claw

With ravine, shriek'd against his creed.

The Complete Poetical Works of Tennyson
In Memoriam A.H.H.
LVI, Stanza IV

Thiery, Paul Henri
The source of man's unhappiness is his ignorance of Nature.

The System of Nature
Volume I
Preface (p. v)

Tobler, Georg Christoph
Nature! We are surrounded and embraced by her—powerless to leave her and powerless to enter her more deeply!

In D. Miller (ed.)
Scientific Studies
Volume 12
Nature (p. 3)

Turgenov, Ivan
However much you knock at nature's door, she will never answer you in comprehensible words because she is dumb. She will utter a musical sound, or a moan like a harp string, but you don't expect a song from her.

On the Eve
Chapter I (p. 10)

Nature is not a temple, but a workshop, and man's the workman in it.

Fathers and Sons
Chapter IX (p. 33)

Twain, Mark
How blind and unreasoning and arbitrary are some of the laws of nature—most of them, in fact!

A Double-Barreled Detective Story
Chapter III (p. 28)

Nature makes the locust with an appetite for crops; man would have made him with an appetite for sand.

Following the Equator
Volume I
Chapter XXX (p. 297)

It is strange and fine—Nature's lavish generosities to her creatures. At least to all of them except man. For those that fly she has provided a home that is nobly spacious—a home which is forty miles deep and envelopes the whole globe, and has not an obstruction in it. For those that swim she has provided a more than imperial domain—a domain which is miles deep and covers four-fifths of the globe. But as for man, she has cut him off with the mere odds and ends of the creation. She has given him the thin skin, the meager skin which is stretched over the remaining one-fifth—the naked bones stick up through it in most places. On the one-half of this domain he can raise snow, ice, sand, rocks, and nothing else. So the valuable part of his inheritance really consists of but a single fifth of the family estate; and out of it he has to grub hard to get enough to keep him alive and provide kings and soldiers and powder to extend the blessings of civilization with. Yet, man, in his simplicity and complacency and inability to cipher, thinks Nature regards him as the important member of the family—in fact, her favorite. Surely, it must occur to even his dull head, sometimes, that she has a curious way of showing it.

Following the Equator
Volume II
Chapter XXVI (p. 311)

Unknown

A simple bard of Nature I
Whose vernal Muse delights to chant
The objects of the earth and sky,
The things that walk, the things that fly
And those that can't.

Source unknown

von Baeyer, Adolf

What makes a great scientist? He must not command but listen; he must adapt himself to what he hears and reshape himself accordingly...The ancient empiricists already did this. They put their ear to Nature. The modern scientist does the same...Coming nearer to Nature has a very special effect on people. They develop very differently from someone who confronts Nature with preconceived ideas. Someone who approaches Nature with set ideas will, so to speak, stand before it like a general. He will want to issue orders to Nature.

In Richard Willstätter
From My Life
Chapter 6 (p. 140)

von Goethe, Johann Wolfgang

Nature! We are surrounded and embraced by her: powerless to separate ourselves from her, and powerless to penetrate beyond her.

Without asking, or warning, she snatches us up into her circling dance, and whirls us on until we are tired, and drop from her arms.

Translated by Thomas Huxley
Nature
Volume 1, November 4, 1869 (p. 9)

Whoever wishes to deny nature as an organ of the divine must begin by denying all revelation.

In D. Miller (ed.)
Scientific Studies
Volume 12
Chapter VIII (p. 303)

It is not easy for us to grasp the vast, the supercolossal, in nature; we have lenses to magnify tiny objects but none to make things smaller. And even for the magnifying glass we need eyes like Carus and Nees to profit intellectually from its use. However, since nature is always the same, whether found in the vast or the small, and every piece of turbid glass produces the same blue as the whole of the atmosphere covering the globe. I think it right to seek out prototypal examples and assemble them before me. Here, then, the enormous is not reduced; it is present within the small, and remains as far beyond our grasp as it was when it dwelt in the infinite.

In D. Miller (ed.)
Scientific Studies
Volume 12
Chapter VIII (p. 304)

Where shall I, endless Nature, seize on thee?

Faust
The First Part, l. 455

When a man of lively intellect first responds to Nature's challenge to be understood, he feels irresistibly tempted to impose his will upon the natural objects he is studying. Before long, however, they close in upon him with such force as to make him realize that he in turn must now acknowledge their might and hold in respect the authority they exert over him.

Botanical Writings
On Morphology
Formation and Transformation (p. 21)

But Nature brooks no foolery; she is always true, always serious, always strict; she is always right, and the mistakes and errors are always ours.

She scourns the inept and submits and reveals her secrets only to the apt, the true, and the pure.

In J.P. Eckermann
Conversations with Goethe
Friday, February 13, 1829 (p. 144)

von Linne, Carl
It is the exclusive property of man, to contemplate and to reason on the great book of nature. She gradually unfolds herself to him, who with patience and perseverance, will search into her mysteries; and when the memory of the present and of past generations shall be obliterated, he shall enjoy the high privilege of living in the minds of his successors, as he has been advanced in the dignity of his nature, by the labours of those who went before him.

Species Plantarum

Walker, John
The objects of nature sedulously examined in their native state, the fields and mountains must be traversed, the woods and waters explored, the ocean must be fathomed and its shores scrutinized by everyone that would become proficient in natural knowledge. The way to knowledge of natural history is to go to the fields, mountains, the oceans, and observe, collect, identify, experiment and study.

Lectures on Geology
Biographical Introduction (p. xvii)

Nature consults no philosophers.

Lectures on Geology
Biographical Introduction (p. xxxi)

Ward, Lester Frank
An entirely new dispensation has been given to the world. All the materials and forces of nature have been thus placed completely under the control of one of the otherwise least powerful of the creatures inhabiting the earth...Nature has thus been made the servant of man.

Glimpses of the Cosmos
Volume III
Mind as a Social Factor (p. 370)

Warner, Charles Dudley
Nature is, in fact, a suggester of uneasiness, a promoter of pilgrimages and of excursions of the fancy which never come to any satisfactory haven.

Backlog Studies
Ninth Study, Section II (p. 203)

Whitehead, Alfred North
Thus we gain from the poets the doctrine that a philosophy of nature must concern itself with at least these five notions: change, value, eternal objects, endurance, organism, interfusion.

Science and the Modern World
The Romantic Reaction (p. 127)

We have to remember that while nature is complex with timeless subtlety, human thought issues from the simple-mindedness of beings whose active life is less than half a century.

An Enquiry Concerning the Principles of Natural Knowledge
Part I
Chapter I, Section 3.8 (p. 15)

Wilde, Oscar
It seems to me that we all look at nature too much, and live with her too little.

De Profundis (p. 158)

And then Nature is so indifferent, so unappreciative. Whenever I am walking in the park here, I always feel that I am no more to her than the cattle that browse on the slope, or the burdock that blooms in the ditch.

Intentions
The Decay of Lying

Nature is so uncomfortable. Grass is hard and lumpy and damp, and full of dreadful insects.

Intentions
The Decay of Lying

Willstätter, Richard
It is the scientist's lot, as it is the artist's, to be less important than his work. He who is chosen to lift the veil from Nature's secrets will be easily overshadowed by the creation he has revealed and which makes him immortal.

From My Life
Chapter 6 (p. 141)

Wordsworth, William
I have learned
To look on nature, not as in the hour
Of thoughtless youth; but hearing oftentimes
The still, sad music of humanity.

The Complete Poetical Works of William Wordsworth
Lines Composed a Few Miles above Tintern Abbey
l. 88–91

Come forth into the light of things;
Let Nature be your teacher.

The Complete Poetical Works of William Wordsworth
The Tables Turned
l. 15–16

Worster, D.

Nature, many have begun to believe, is fundamentally erratic, discontinuous, and unpredictable. It is full of seemingly random events that elude our models of how things are supposed to work. As a result, the unexpected keeps hitting us in the face. Clouds collect and disperse, rain falls or doesn't fall, disregarding our careful weather predictions, and we cannot explain why. A man's heart beats regularly year after year, then abruptly begins to skip a beat now and then. Each little snowflake falling out of the sky turns out to be completely unlike any other. If the ultimate test of any body of scientific knowledge is its ability to predict events, then all of the sciences...—physics, chemistry, climatology, economics, ecology—fail the test regularly. They all have been announcing laws, designing models, predicting what an individual atom or person is supposed to do; and now, increasingly, they are beginning to confess that the world never quite behaves the way it is supposed to do.

Environmental History Review
The Ecology of Chaos and Harmony (p. 13)
Volume 14, 1990

Yogananda, Paramahansa

Because modern science tells us how to utilize the powers of Nature, we fail to comprehend the Great Life in back of all names and forms. Familiarity with Nature has bred a contempt for her ultimate secrets; our relation with her is one of practical business. We tease her, so to speak, to discover the ways in which she may be forced to serve our purposes; we make use of her energies, whose Source yet remains unknown. In science our relation with Nature is like that between an arrogant man and his servant; or, in a philosophical sense, Nature is like a captive in the witness box. We cross-examine her, challenge her, and minutely weigh her evidence in human scales that cannot measure her hidden values.

Autobiography of a Yogi
Chapter 35 (p. 337–8)

OBSERVATION

Altmann, Jeanne
The true situation may be the opposite of the apparent one.

Baboon Mothers and Infants
Chapter 9 (p. 169)

Burroughs, John
Unadulterated, unsweetened observations are what the real nature-lover craves. No man can invent incidents and traits as interesting as the reality.

Ways of Nature
Ways of Nature (p. 15)

Darwin, Charles
I have an old belief that a good observer really means a good theorist.

In Francis Darwin (ed.)
More Letters of Charles Darwin
Letter to Bates
22 November, 1860 (p. 195)

Let theory guide your observations, but till your reputation is well established, be sparing in publishing theory. It makes persons doubt your observations.

In Francis Darwin (ed.)
More Letters of Charles Darwin
Volume II
Darwin to Scott
June 6, 1863 (p. 323)

I am a firm believer that without speculation there is no good and original observation.

In Francis Darwin (ed.)
The Life and Letters of Charles Darwin
Volume I
To Wallace
December 22, 1857 (p. 465)

Drake, Daniel
If observation be the soil, reading is the manure of intellectual culture.

Introductory Lecture, on the Means of Promoting the Intellectual
Improvement of Students and Physicians of the Mississippi Valley (p. 16)

Eddington, Sir Arthur Stanley
Let us suppose that an ichthyologist is exploring the life of the ocean. He casts a net into the water and brings up a fishy assortment. Surveying his catch, he proceeds in the usual manner of a scientist to systematise what it reveals... In applying this analogy, the catch stands for the body of knowledge which constitutes physical science, and the net for the sensory and intellectual equipment which we use in obtaining it. The casting of the net corresponds to observation; for knowledge which has not been or could not be obtained by observation is not admitted into physical science.

The Philosophy of Physical Science
Chapter II, Section I (p. 16)

Emerson, Ralph Waldo
The difference between landscape and landscape is small but there is a great difference in the beholders.

Essays
Second Series
Nature (p. 170)

Grew, Nehemiah
If...an inquiry into the Nature of *Vegetation* may be of good Impart; It will be requisite to see, first of all, What may offer it self to be enquired of; or to understand, what or *Scope* is: That so doing, we may take our aim the better in making, and having made, in applying our Observations thereunto.

The Anatomy of Plants
An Idea of a Philosophical History of Plants (p. 3)

Hales, Stephen
...it is from long experience chiefly that we are to expect the most certain rules of practice, yet it is withal to be remembered, that observations, and to put us upon the most probable means of improving any art, is to get the best insight we can into the nature and properties of those things which we are desirous to cultivate and improve.

Vegetable Staticks
The Conclusion (p. 214)

Hanson, Norwood Russell
The observer may not know what he is seeing: he aims only to get his observations to cohere against the background of established knowledge. This seeing is the goal of observation.

Patterns of Discovery
Chapter I (p. 20)

Liebig, Justus
However numerous our observations may be, yet, if they only bear on one side of a question, they will never enable us to penetrate the essence of a natural phenomenon in its full significance.

Animal Chemistry
Preface (p. xxxii)

Minnaert, M.
It is indeed wrong to think that the poetry of Nature's moods in all their infinite variety is lost on one who observes them scientifically, for the habit of observation refines our sense of beauty and adds a brighter hue to the richly coloured background against which each separate fact is outlined. The connection between events, the relation of cause and effect in different parts of a landscape, unite harmoniously what would otherwise be merely a series of detached sciences.

The Nature of Light and Colour in the Open Air
Preface (p. v)

Simpson, George Gaylord
It is inherent in any definition of science that statements that cannot be checked by observation are not really about anything... or at the very best, they are not science.

Science
The Non-prevalence of Humanoids (p. 769)
Volume 143, 1964

Steinbeck, John
...one can live in a prefabricated world, smugly and without question, or one can indulge perhaps the greatest human excitement: that of observation to speculation to hypothesis. This is a creative process, probably the highest and most satisfactory we know.

In Edward F. Ricketts, Jack Calvin and Joel W. Hedgpeth
Between Pacific Tides
Prefaces (p. xi)

There are good things to see in the tidepools and there are exciting and interesting thoughts to be generated from the seeing. Every new eye applied to the peep hole which looks out at the world may fish in some new beauty and some new pattern, and the world of the human mind must be enriched by such fishing.

In Edward F. Ricketts, Jack Calvin, and Joel W. Hedgpeth
Between Pacific Tides
Prefaces (p. xi)

Sterne, Laurence

What a large volume of adventures may be grasped within this little span of life by him who interests his heart in everything and who, having eyes to see what time and chance are perpetually holding out to him as he journeyeth on his way, misses nothing he can *fairly* lay his hands on.

A Sentimental Journey
In the Street

Teale, Edwin Way

For observing nature, the best pace is a snail's pace.

Circle of the Seasons
July 14 (p. 150)

Thomas, Lewis

The role played by the observer in biological research is complicated but not bizarre: he or she simply observes, describes, interprets, maybe once in a while emits a hoarse shout, but that is that; the act of observing does not alter fundamental aspects of the things observed, or anyway isn't supposed to.

The Medusa and the Snail
An Apology (p. 88)

Whitehead, Alfred North

We habitually observe by the method of difference. Sometimes we see an elephant, and sometimes we do not. The result is that an elephant, when present, is noticed.

Process and Reality
Chapter I, Section II (p. 6)

Wright, R.D.

Whatever happened to the terms *probability* and *observation?* Are statements of high probability now to be deified by calling them *truths?* Does a set of consistent observations become *fact?* When I teach biology to the college student, the nature of information mandates that the class and I preserve a healthy skepticism regarding both the broad generalizations and the specific statements of the discipline. Fact and truth are terms we almost never use. There is nothing shameful in describing what we know as having a certain probability, following from observations that have a degree of imprecision. That's the nature of science, including the science of evolution.

Bioscience
Letters (p. 788)
Volume 31, Number 11, December 1981

OCCAM'S RAZOR

Crick, Francis Harry Compton
While Occam's razor is a useful tool in the physical sciences, it can be a very dangerous implement in biology. It is thus very rash to use simplicity and elegance as a guide in biological research.

What Mad Pursuit
Chapter 13 (p. 138)

OCEAN

Beston, Henry

The seas are the heart's blood of the earth.

The Outermost House
The Headlong Wave (p. 47)

Carson, Rachel

For all at last return to the sea—to Oceanus, the ocean river, like the ever-flowing stream of time, the beginning and the end.

The Sea Around Us
The Encircling Sea (p. 209)

The edge of the sea is a strange and beautiful place. All through the long history of Earth it has been an area of unrest where waves have broken heavily against the land, where the tides have pressed forward over the continents, receded, and then returned. For no two successive days is the shore line precisely the same. Not only do the tides advance and retreat in their eternal rhythms, but the level of the sea itself is never at rest. It rises or falls as the glaciers melt or grow, as the floor of the deep ocean basins shift under its increasing load of sediments, or as the earth's crust along the continental margins warps up or down in adjustment to strain and tension. Today a little more land may belong to the sea, tomorrow a little less. Always the edge of the sea remains an elusive and indefinable boundary.

The Edge of the Sea
The Marginal World (p. 1)

Forbes, Edward

...beneath the waves there are many dominions yet to be visited, and kingdoms to be discovered; and he who venturously brings up from the abyss enough of their inhabitants to display the physiognomy of the country, will taste that cup of delight, the sweetness of whose draught those only who have made a discovery know.

The Natural History of the European Seas
Chapter I (p. 11)

Hardy, Thomas

Who can say of a particular sea that it is old? Distilled by the sun, kneaded by the moon, it is renewed in a year, in a day, or in an hour.

The Return of the Native
Chapter I (p. 7)

Henderson, Lawrence

No philosopher's or poet's fancy, no myth of a primitive people has ever exaggerated the importance, the usefulness, and above all the marvelous beneficence of the ocean for the community of living things.

The Fitness of the Environment
Chapter V (p. 190)

Horsfield, Brenda
Stone, Peter Bennet

...there is on the other hand some encouragement in the reflection that Oceanography has usually only ruined the reputations of people who dared to speculate too little and thought on too small a scale. She has smiled most benignly on those who backed the most daring and outrageous possibility...

The Great Ocean Business
Chapter 7 (p. 150)

Mishima, Yukio

Down beneath the spray, down beneath the whitecaps, that beat themselves to pieces against the prow, there were jet-black invisible waves, twisting and coiling their bodies. They kept repeating their patternless movements, concealing their incoherent and perilous whims.

The Sound of Waves
Chapter 14 (p. 125)

Ovid

The face of places, and their forms decay;
And that is solid Earth, that once was sea:
Seas in their turn retreating from the shore,
Make solid land, what ocean was before...

In S. Garth (ed.)
Ovid's Metamorphoses, in Fifteen Books
Metamorphoses
15th Book (p. 496)

Spenser, Edmund
For all, that here on earth we dreadfull hold,
Be but as bugs to fearen babes withall,
Compared to the creatures in the seas entrall.

The Complete Poetical Works of Edmund Spenser
The Faerie Queene
Book II, Canto XII, Stanza XXV

Whitman, Walt
To me the sea is a continual miracle,
The fishes that swim—the rocks—the motion of the waves—the ships with men in them,
What stranger miracles are there?

Complete Poetry and Collected Prose
Miracles

ORGANIC

Darwin, Charles
An organic being is a microcosm—a little universe, formed of a host of self-propagating organisms, inconceivably minute and numerous as the stars of heaven.

In John Lubbock
The Beauties of Nature (p. 72)

Reichenbach, Hans
. . . whereas inorganic nature was seen to be controlled by the laws of cause and effect, organic nature appeared to be governed by the law of purpose and means.

The Rise of Scientific Philosophy
Chapter 12 (p. 192)

von Schubert, G.H.
In the form of the organic world nature rises again from the grave of decay, and the cause of the organic inceptions has been simultaneously that of the decline of the inorganic world. Thus a new period is merrily built on top of the ruins of the old submerged one, in the hope of establishing its handiwork more firmly on the deep foundations of the most remote times, not as a result of the permanence of corporeal mass, but through spiritual strength.

Ansichten von der Nachtseite der Naturwissenschaften (p. 198)

ORGANISM

Evans, Howard Ensign

It has been said that for every problem concerning living things there is an organism ideal for its solution. It is probable that there are still undiscovered species living that hold the answers to problems that face us now or will in the future.

Pioneer Naturalist: The Discovery and Naming of North American Plants and Animals
Naturalists, Then and Now (p. 267)

Jacob, François

And one of the deepest, one of the most general functions of living organisms is to look ahead, to produce future as Paul Valéry put it.

The Possible and the Actual
Time and the Invention of the Future (p. 66)

Jones, J.S.
Ebert, D.
Stearns, S.C.

No organism can do everything. Every creature is restricted by constraints of various kinds. Many of these arise from the facts of history and the nature of evolution, both of which can proceed only from where they left off.

In R.J. Berry, T.J. Crawford and G.M. Hewitt (eds)
Genes in Ecology
Life History and Mechanical Constraints on
Reproduction in Genes, Cells and Waterfleas (p. 393)

Unknown

Under the most rigorously controlled conditions of pressure, temperature, volume, humidity, and other variables the organism will do as it pleases.

Source Unknown

von Goethe, Johann Wolfgang
Basic characteristics of an individual organism: to divide, to unite, to merge into the universal, to abide in the particular, to transform itself, to define itself, and as living things tend to appear under a thousand conditions, to arise and vanish, to solidify and melt, to freeze and flow, to expand and contract. Since these effects occur together, any or all may occur at the same moment.

In D. Miller (ed.)
Scientific Studies
Volume 12
Chapter VIII (pp. 303–4)

ORGANIZATION

Eiseley, Loren
Men talk much of matter and energy, of the struggle for existence that molds the shape of life. These things exist, it is true; but more delicate, elusive, quicker than fins in water, is that mysterious principle known as "organization," which leaves all other mysteries concerned with life stale and insignificant by comparison. For that without organization life does not persist is obvious. Yet this organization itself is not strictly the product of life, nor of selection. Like some dark and passing shadow within matter, it cups out the eyes' small windows or spaces the notes of a meadow lark's song in the interior of a mottled egg.

The Immense Journey
The Flow of the River (p. 26)

Kauffman, Stuart
If biologists have ignored self-organization, it is not because self-ordering is not pervasive and profound. It is because we biologists have yet to understand how to think about systems governed simultaneously by two sources of order. Yet who seeing the snowflake, who seeing simple lipid molecules cast adrift in water forming themselves into cell-like hollow lipid vesicles, who seeing the potential for the crystallization of life in swarms of reacting molecules, who seeing the stunning order for free in networks linking tens upon tens of thousands of variables, can fail to entertain a central thought: if ever we are to attain a final theory in biology, we will surely, surely have to understand the commingling of self-organization and selection. We will have to see that we are the natural expressions of a deeper order. Ultimately, we will discover in our creation myth that we are expected after all.

At Home in the Universe (p. 112)

Needham, Joseph
Organization is not something mystical and inaccessible to scientific attack, but rather the basic problem confronting the biologist... It is for

us to investigate the nature of this biological organization, not to abandon it to the metaphysicians because the rules of physics do not seem to apply to it.

Order and Life
Chapter I (pp. 7, 17–18)

Organization and Energy are the two fundamental problems which all science has to solve.

Time: The Refreshing River (p. 33)

Simpson, George Gaylord

The point about explanation in biology that I would particularly like to stress is this: to understand organisms one must explain their organization. It is elementary that one must know what is organized and how it is organized, but that does not explain the fact or the nature of the organization itself. Such explanation requires knowledge of how an organism came to be organized and what function the organization serves. Ultimate explanation in biology is therefore necessarily evolutionary.

This View of Life: The World of an Evolutionist
Chapter 6 (p. 113)

Szent-Györgyi, Albert

One of the most basic principles of biology is organization, which means that two things put together in a specific way form a new unit, a system, the properties of which are not additive and cannot be described in terms of the properties of the constituents. As points may be connected to letters, letters to words, words to sentences, etc., so atoms can join to molecules, molecules to organelles, organelles to cells, etc., every level of organization having a new meaning of its own and offering exciting vistas and possibilities.

Bioenergetics
Chapter 6 (p. 39)

Woodger, Joseph Henry

If the concept of organization is of such importance as it appears to be it is something of a scandal that we have no adequate conception of it. The first duty of the biologist would seem to be to try and make clear this important concept. Some biochemists and physiologists...express themselves as though they really believed that if they concocted a mixture with the same chemical composition as what they call 'protoplasm' it would proceed to 'come to life.' This is the kind of nonsense which results from forgetting or being ignorant of organization.

Biological Principles (p. 291)

ORIGINS

de Chardin, Teilhard
To push anything back into the past is equivalent to reducing it to its simplest element. Traced as far as possible in the direction of their origins, the last fibers of the human aggregate are lost to view and are merged in our eyes with the very stuff of the universe.

The Phenomenon of Man
Book 1, Chapter I (p. 39)

ORNITHOLOGY

Unknown

...the philosophy of science is just about as useful to scientists as ornithology is to birds.

In S. Weinberg
Nature
Newtonianism, Reductionism and the Art of Congressional Testimony
Volume 330, Number 6147, 3–9 December 1987 (p. 433)

Vidal, Gore

To a man, ornithologists are tall, slender, and bearded so that they can stand motionless for hours, imitating kindly trees, as they watch for birds.

Armageddon? Essays 1983–1987
Mongolia (p. 131)

PARASITE

Frost, Robert
Will the blight end the chestnut?
The farmers rather guess not.
It keeps smoldering at the roots
And sending up new shoots
Till another parasite
Shall come to end the blight.

The Poetry of Robert Frost
Evil Tendencies Cancel

Shakespeare, William
... unbidden guests
Are often welcomest when they are gone.

The First Part of King Henry the Sixth
Act II, Scene II, l. 55–6

PATTERNS

Derry, Gregory N.
In trying to understand nature, we rarely attempt to grasp completely every possible detail. If we did, we'd be overwhelmed by the mass of inconsequential information. As a result, we would miss the truly interesting patterns and relationships that give us scientific insight.

What Science is and How it Works
Chapter 6 (p. 69)

Flannery, Maura C.
The patterns and rhythms of nature, science as a search for order, form as a central problem in biology—are rarely emphasized in research reports and in texts, they are nevertheless powerful concepts that direct and inform biologists' work.

Perspectives in Biology and Medicine
Biology is Beautiful
Volume 35, Number 3, Spring 1992 (p. 426)

Huxley, Aldous
The difference between a piece of stone and an atom is that an atom is highly organised, whereas the stone is not. The atom is a pattern, and the molecule is a pattern, and the crystal is a pattern; but the stone, although it is made up of these patterns, is just a mere confusion. It's only when life appears that you begin to get organisation on a larger scale. Life takes the atoms and molecules and crystals; but, instead of making a mess of them like the stone, it combines them into new and more elaborate patterns of its own.

Time Must Have a Stop
Chapter XIV (p. 145)

Lowell, Amy
Christ! What are patterns for?

The Complete Poetical Works of Amy Lowell
Patterns

MacArthur, Robert H.
To do science is to search for repeated patterns, not simply to accumulate facts, and to do the science of geographical ecology is to search for patterns of plants and animal life that can be put on a map.

Geographical Ecology
Introduction (p. 1)

PERCEPTION

Whitehead, Alfred North
Our problem is, in fact, to fit the world to our perceptions, and not our perceptions to the world.

The Organisation of Thought
Chapter VIII (p. 228)

PHOTOSYNTHESIS

Baum, Harold
When sunlight bathes the chloroplast, and photons are absorbed
The energy's transduced so fast that food is quickly stored,
Photosynthetic greenery traps light the spectrum through
Then dark pathway machinery fixes the CO_2.

The Biochemists' Handbook
Photosynthesis
(Tune: "Auld Lang Syne")

Pallister, William
The sunlight gives the stimulus
Which makes a plant of you;
Your chemic process puzzles us,
We look and see you do
Your photo-synthesis, and thus
Grow and divide in two.

Poems of Science
The Nature of Things
Euglena viridis (p. 5)

Rabinowitch, E.I.
In photosynthesis we are like travelers in an unknown country around whom the early morning fog slowly begins to rise, vaguely revealing the outlines of the landscape. It will be thrilling to see it in bright daylight!

In A Scientific American Book
The Physics and Chemistry of Life
Photosynthesis (p. 47)

PLANTS

Aristotle

...next after lifeless things in the upward scale comes the plant, and of plants one will differ from another as to its amount of apparent vitality; and, in a word, the whole genus of plants, whilst it is devoid of life as compared with an animal, is endowed with life as compared with other corporeal entities.

History of Animals
Book VIII, I, 588b[5]

Borland, Hal

There are no idealists in the plant world and no compassion. The rose and the morning glory know mercy. Bindweed, the morninglory, will quickly choke its competitors to death, and the fencerow rose will just as quietly crowd out any other plant that tried to share its roothold. Idealism and mercy are human terms and human concepts.

Book of Days
22 July 1976 (pp. 188–9)

Cvikota, Raymond J.

Plant food. Bud plasma.

Quote
October 13, 1968 (p. 317)

Emerson, Ralph Waldo

Plants are the young of the world, vessels of health and vigor; but they grope ever upwards towards consciousness; the trees are imperfect men, and seem to bemoan their imprisonment, rooted in the ground.

Essays
Second Edition
Nature (p. 174)

Gerhard, John
Among the manifold creatures of God (right Honorable, and my singular good Lord) that have all in all ages diversely entertained many excellent wits, and drawne them to the contemplation of the divine wisdome, none have provoked mens studies more, or satisfied their desires so much as Plants have done, and that upon Just and worthy causes: For if delight may provoke mens labor, what greater delight is there than to behold the earth apparelled with plants, as with a robe of embroidered worke, set with Orient pearles, and garnished with great diversitie of rare and costly jewels?

The Herball or Generall Historie of Plantes (p. 4)

Haldane, J.B.S.
The simplest plants, such as the green algae growing in stagnant water or on the bark of trees, are mere round cells. The higher plants increase their surface by putting out leaves and roots. Comparative anatomy is largely the story of the struggle to increase surface in proportion to volume.

In J.R. Newman (ed.)
The World of Mathematics (p. 954)

Turner, William
Although (most mighty and Christian Prince) there be many noble and excellent arts and sciences, which no man doubteth, but that almighty God the author of all goodness hath given unto us by the hands of the heathen, as necessary unto the use of mankind, yet is there none among them all which is so openly commended by the verdict of any holy writer in the Bible, as is the knowledge of plants, herbs and trees...

In George T.L. Chapman and Marilyn N. Tweddle (eds)
A New Herball
Part I (p. 213)

von Goethe, Johann Wolfgang
The primordial plant is turning out to be the most marvelous creation in the world, and nature itself will envy me because of it. With this model and the key to it an infinite number of plants can be invented, which must be logical, that is, if they do not exist, they *could* exist, and are not mere artistic or poetic shadows and semblances, but have an inner truth and necessity.

Italian Journey
Naples, 17 May, 1787 (p. 255)

We will see the entire plant world, for example, as a vast sea which is as necessary to the existence of individual insects as the oceans and rivers are to the existence of individual fish, and we will observe that an enormous number of living creatures are born and nourished in this ocean of plants.

Ultimately we will see the whole world of animals as a great element in which one species is created, or at least sustained, by and through another. We will no longer think of connections and relationships in terms of purpose or intention. This is the only road to progress in understanding how nature expresses itself from all quarters and in all directions as it goes about its work of creation.

In D. Miller (ed.)
Scientific Studies
Volume 12
Chapter II (p. 55)

Anyone who pays a little attention to the growth of plants will readily observe that certain of their external members are sometimes transformed so that they assume—either wholly or in some lesser degree—the form of the members nearest in the series. Thus, for example, the usual process by which a single flower becomes double, is that, instead of filaments and anthers, petals are developed; these either show a complete resemblance in form and color to the other leaves of the corolla, or they still carry some visible traces of the origin.

If we note that it is in this way possible for the plant to take a step backwards and thus to reverse the order of growth, we shall obtain so much the more insight into Nature's regular procedure; and we shall make the acquaintance of the laws of transmutation, according to which she produces one part from another, and sets before us the most varied forms through modification of a single organ.

Chronica Botanica
An Attempt to Interpret the Metamorphosis of Plants
Introduction
Section 1 and Section 3 (p. 91)
Volume 10, Number 2, Summer 1946

von Linne, Carl
For wealth disappears, the most magnificent houses fall into decay, the most numerous family at some time or another comes to an end: the greatest and the most prosperous kingdoms can be overthrown: but the whole of Nature must be blotted out before the race of plants passes away, and he is forgotten who in Botany held up the torch.

Critica Botanica
Generic Names (p. 68)

Wisdom of Solomon 7:20
To know…the diversities of plants, and the virtues of roots.

The Bible

POLLUTION

Carson, Rachel
These sprays, dusts, and aerosols are now applied almost universally to farms, gardens, forests, and homes—nonselective chemicals that have the power to kill every insect, the "good" and the "bad," to still the song of birds and the leaping of fish in the streams, to coat the leaves with a deadly film, and to linger on in soil—all this though the intended target may be only a few weeds or insects. Can anyone believe it is possible to lay down such a barrage of poisons on the surface of the earth without making it unfit for all life? They should not be called "insecticides," but "biocides."

Silent Spring
Chapter 2 (pp. 7–8)

Eliot, T.S.
There are flood and drouth
Over the eyes and in the mouth,
Dead water and dead sand
Contending for the upper hand.
The parched eviscerate soil
Gapes at the vanity of toil,
Laughs without mirth.
This is the death of earth.

Four Quartets
Little Gidding
Part II, stanza 2

Peacock, Thomas Love
... they have poisoned the Thames and killed the fish in the river. A little further development of the same wisdom and science will complete the poisoning of the air, and kill the dwellers on the banks.

Gryll Grange
Misnomers (p. 11)

Shakespeare, William

...this most excellent canopy, the air, look you, this brave o'erhanging firmament, this majestical roof fretted with golden fire, why, it appears no other thing to me than a foul and pestilent congregation of vapours.

Hamlet, Prince of Denmark
Act II, Scene II, l. 311–15

Taylor, John

Then by the Lords Commissioners, and also
By my good King (whom all true subjects call so),
I was commanded with the Water Baylie,
To see the rivers cleaned, both night and dayly.
Dead Hogges, Dogges, Cates and well flayed Carryon Horses,
Their Noysom Corpses soyled the Water Courses;
Both Swines' and Stable dynge, beasts guts and garbage,
Street dirt, with Gardners' Weeds and Rotten Herbage.
And from those Waters' filthy putrification
Our Meat and Drinke were made, which bred Infection.
Myself and partner, with cost paines and Travell,
Saw all made clean, from Carryon, Mud and Gravell,
And now and then was punisht a Delinquent,
By which good meanes away the filth and stink went.

Unknown
The American Biology Teacher
An Echo from the Past (p. 208)
Volume 35, Number 4, April 1973

Toffler, Alvin

...industrial vomit...fills our skies and seas. Pesticides and herbicides filter into our foods. Twisted automobile carcasses, aluminum cans, non-returnable glass bottles and synthetic plastics form immense kitchen middens in our midst as more and more of our detritus resists decay. We do not even begin to know what to do with our radioactive wastes—whether to pump them into the earth, shoot them into outer space, or pour them into the oceans. Our technological powers increase, but the side effects and potential hazards also escalate.

Future Shock
Chapter 19 (p. 429)

PRAYER

Stephens, James
Little things, that run, and quail,
And die, in silence and despair!

Little things, that fight, and fail,
And fall, on sea, and earth, and air!

All trapped and frightened little things,
The mouse, the coney, hear our prayer!

As we forgive those done to us,
—The lamb, the linnet, and the hare—

Forgive us all our trespasses,
Little creatures, everywhere!

Collected Poems
Little Things

Wilbur, Richard
When I must come to you, O my God, I pray
It be some dusty-roaded holiday,
And even as in my travels here below,
I beg to choose by what road I shall go
To Paradise, where the clear stars shine by day.
I'll take a walking-stick and go my way,
And to my friends the donkeys I shall say,
"I am Francis Jammes, and I'm going to Paradise,
For there is no hell in the land of the loving God."
And I'll say to them: "Come sweet friends of the blue skies,
Poor creatures who with a flap of the ears or a nod
Of the head shake off the buffets, the bees, the flies..."

Let me come with these donkeys, Lord, into your land,
These beasts who bow their heads so gently, and stand

373

With their small feet joined together in a fashion
Utterly gentle, asking your compassion.
I shall arrive, followed by their thousands of ears,
Followed by those with baskets, at their flanks,
By those who lug the carts of mountebanks
Or loads of feather-dusters and kitchen-wares,
By those with humps of battered water-cans,
By bottle-shaped she-asses who halt and stumble,
By those tricked out in little pantaloons
To cover their wet, blue galls where flies assemble
In whirling swarms, making a drunken hum.
Dear God, let it be with these donkeys that I come,
And let it be that angels lead us in peace
To leafy streams where cherries tremble in air,
Sleek as the laughing flesh of girls; and there
In that heaven of souls let it be that, Leaning above
Your divine waters, I shall resemble these donkeys,
Whose humble and sweet poverty will appear
Clear in the clearness of your eternal love.

Things of This World
Francis Jammes: A Prayer to go to Paradise with the Donkeys

PRIMORDIAL

de Maupassant, Guy

Nothing is more disturbing, nothing, more disquieting, more terrifying occasionally, than a fen. Why should this terror hang over these low plains covered with water? Is it the vague rustling of the rushes, the strange Will-o'-the-wisps, the profound silence which envelops them on calm nights, or is it the strange mists, which hang over the rushes like a shroud; or else it is the imperceptible splashing, so slight and so gentle, and sometimes more terrifying than the cannons of men of the thunders of skies, which make these marshes resemble countries which none has dreamed of, terrible countries concealing an unknown and dangerous secret.

No, something else belongs to it, another mystery, more profound and graver floats amid these thick mists, perhaps the mystery of the creation itself! For was it not in stagnant and muddy water, amid the heavy humidity of moist land under the heat of the sun, that the first germ of life vibrated and expanded to the day?

Love (p. 264)

Newman, Joseph S.

A highly speculative void
Divides the germ and anthropoid
But we've discovered certain clues
In fossilized primordial ooze
Where ancient polyps lived and died
And countless myriads multiplied.

Poems for Penguins
Biology

Shakespeare, William

In the cauldron boil and bake;
Eye of newt and toe of frog,
Wool of bat and tongue of dog,

375

Adder's fork and blind-worm's sting,
Lizard's leg and howlet's wing...

Macbeth
Act IV, Scene I, l. 13–17

PROTEIN

Brenner, Sydney
Nobody publishes theory in biology—with few exceptions. Instead, they get out the structure of still another protein.

In H.F. Judson
The Eighth Day of Creation
Chapter 4 (p. 218)

Mulder, Gerard Johannes
There is present in plants and in animals a substance which...is without a doubt the most important of all the known substances in living matter, and, without it, life would be impossible on our planet. This material has been named Protein.

In A Scientific American Book
The Physics and Chemistry of Life
Proteins (p. 58)

Sherrington, Sir Charles
"Life" is a maker of proteins.

Man on His Nature
Chapter III (p. 81)

PROTOZOA

Pallister, William
PROTOZOA, five thousand, each species minute
And quite simple in structure, the first to appear,
The ancestors of races, arrived and *en route*;
All of these live in water, not in atmosphere;
Subdividing in billions, amoebae and germs,
They permit of the study of life in small terms.

Poems of Science
Beginnings
Animal Life (p. 139)

REPRODUCTION

Ecclesiastes 1:2
One generation passeth away, and another generation cometh.

The Bible

Fletcher, Joseph
Our basic ethical choice as we consider man's new control over himself, over his body and his mind as well as over his society and environment, is still what it was when primitive men holed up in caves and made fires. Chance versus control. Should we leave the fruits of human reproduction to take shape at random, keeping our children dependent on accidents of romance and genetic endowment, of sexual lottery or what one physician calls "the meiotic roulette of his parents' chromosomes?" Or should we be responsible about it, that is, exercise our rational and human choice, no longer submissively trusting to the blind worship of raw nature?

The Ethics of Genetic Control
Chapter I, Trying to be Natural (p. 36)

Genesis 8:17
That they may breed abundantly in the earth, and be fruitful, and multiply upon the earth.

The Bible

Walters, Mark Jerome
Courtship is the bringing together of individuals. Conception is the bringing together of gametes.

The Dance of Life
Chapter 1 (p. 12)

Zihlman, Adriene

As with most things in life, the debate centers on two themes: food and sex; or to give it a proper academic tone: diet and reproduction.

Yearbook of Physical Anthropology
Sex, Sexes and Sexism in Human Origins (p. 11)
Volume 30, 12 April 1985

REPTILE

ALLIGATOR

Ackerman, Diane
Nothing looks more contented than a resting alligator. The mouth falls naturally into a crumpled smile, the eyes half close in a sleepy sort of way...

The Moon By Whale Light
Chapter 2 (p. 60)

ASP

Flaubert, Gustave
Asp: Animal known through Cleopatra's basket of figs.

Dictionary of Accepted Ideas

CHAMELEON

Wells, Carolyn
The true Chameleon is small,
A lizard sort of thing;
He asn't any ears at all,
And not a single wing.
If there is nothing on the tree,
'Tis the Chameleon you see.

Baubles
How to Tell the Wild Animals

COBRA

Nash, Ogden
This creature fills its mouth with venum
And walks upon its duodenum.
He who attempts to tease the cobra
Is soon a sadder he, and sobra.

Verses from 1929 On
The Cobra

CROCODILE

Carroll, Lewis
How cheerfully he seems to grin,
How neatly spreads his claws,
And welcomes little fishes in
With gently smiling jaws!

The Complete Works of Lewis Carroll
Alice's Adventures in Wonderland
Chapter II (p. 29)

LIZARD

Gardner, John
The Lizard is a timid thing
That cannot dance or fly or sing;
He hunts for bugs beneath the floor
And longs to be a dinosaur.

A Child's Bestiary
The Lizard

Lawrence, D.H.
A lizard ran out on a rock and looked up, listening
no doubt to the sounding of spheres.
And what a dandy fellow! The right toss of a chin for you
And swirl of a tail!

If men were as much men as lizards are lizards
they'd be worth looking at.

The Complete Poems of D.H. Lawrence
Volume I
The Lizard

PYTHON

Nash, Ogden
The python has, and I fib no fibs,
318 pairs of ribs.
In stating this I place reliance
On a séance with one who died for science.
This figure is sworn to and attested;
He counted them while being digested.

Verses from 1929 On
The Python

Prelutsky, Jack
A puzzled python shook its head
and said, "I simply fail
to tell if I am purely neck,
or else entirely tail."

A Pizza the Size of the Sun
A Puzzled Python

RATTLESNAKE

Muir, John
Poor creatures, loved only by their Maker, they are timid and bashful, as mountaineers know, and though perhaps not possessed of much of that charity that suffers long and is kind, seldom, either by mistake or by mishap, do harm to any one. Certainly they cause not the hundredth part of the pain and death that follow the footsteps of the admired Rocky Mountain trapper. Nevertheless, again and again, in season and out of season, the question comes up, "What are rattlesnakes good for?" As if nothing that does not obviously make for the benefit of man has any right to exist; as if our ways were God's ways. Long ago, an Indian to whom a French traveler put this old question replied that their tails were good for toothache, and their heads for fever. Anyhow, they are all, head and tail, good for themselves, and we need not begrudge them their share of life.

In Sally M. Miller (ed.)
John Muir: Life and Work
Part III, Chapter 5 (p. 108)

SNAKE

Aristotle

...in the serpents...[the tongue] is long and forked...For by this arrangement they derive a twofold pleasure from savours, their gustatory sensation being as it were doubled.

On the Parts of Animals
Book II, Chapter 17, 660b

TURTLE

Rudloe, Jack

The timeless turtle will look on as man works feverishly to develop destructive nuclear weapons that will blow the world apart many times over. And perhaps one day when he pops his head up from the sea, he'll see a world empty of man, with barnacles growing on the ruins of the cities and buildings. And somewhere, perhaps on a Mexican beach, a handful of Kemp's ridleys filled with eggs will crawl out on the sand, unmolested and free.

Time of the Turtle
Chapter 9 (p. 106)

RESEARCH

Cussler, Clive
Dirgo, Craig
Research is the key. You can never do enough research. This is so vital I'll repeat it. *You can never to enough research.* . . . Research can either lower the odds or tell you it's hopeless.

The Sea Hunters
Introduction (p. 28)

SCENERY

Darwin, Charles

But there is a growing pleasure in comparing the scenery in different countries, which to a certain degree is distinct from merely admiring its beauty. It depends chiefly on an acquaintance with the individual parts of each view: I am strongly induced to believe that, as in music, the person who understands every note will, if he also possesses a proper taste, more thoroughly enjoy the whole, so he who examines each part of a fine view, may also comprehend the full and combined effect.

The Voyage of the Beagle
Chapter XXI (p. 505)

SCIENCE

Bernard, Claude
...my idea of the science of life, I should say that it is a superb and dramatically lighted hall which may be reached only by passing through a long and ghastly kitchen.

An Introduction to the Study of Experimental Medicine
Part I, Chapter III (p. 15)

Carlyle, Thomas
This world, after all our science and sciences, is still a miracle; wonderful, inscrutable, *magical* and more, to whosoever will *think* of it.

Sartor Restarus & On Heroes
Hero Worship
Lecture I (p. 246)

Carson, Rachel
We live in a scientific age, yet we assume that knowledge of science is the prerogative of only a small number of human beings, isolated and priestlike in their laboratories. This is not true. The materials of science are the materials of life itself. Science is part of the reality of living; it is the what, the how and the why of everything in our experience.

In Paul Brooks
The House of Life: Rachel Carson at Work
Fame (p. 128)

There is one quality that characterizes all of us who deal with the science of the earth and its life—we are never bored.

In Paul Brooks
The House of Life: Rachel Carson at Work
The Closing Journey (p. 324)

Douglas, M.
Wildavsky, A.
In our modern world people are supposed to live and die subject to known, measurable natural forces, not subject to mysterious moral

agencies. That mode of reasoning, indeed, is what makes modern man modern. Science wrought this change between us and nonmoderns. It is hardly true, however, that their universe is more unknown than ours. For anyone disposed to worry about the unknown, science has actually expanded the universe about which we cannot speak with confidence...This is the double-edge thrust of science, generating new ignorance with new knowledge. The same ability to detect causes and connections or parts per trillion can leave more unexplained than was left by cruder measuring instruments.

> *Risk and Culture: An Essay on the Selection of Technical and Environmental Dangers*
> Chapter III (p. 49)

Emerson, Ralph Waldo

The motive of science was the extension of man, on all sides, into Nature, till his hands should touch the stars, his eyes see through the earth, his ears understand the language of beast and bird, and the sense of the wind; and, through his sympathy, heaven and earth should talk with him. But that is not our science.

> *The Conduct of Life*
> Beauty (p. 249)

Empirical science is apt to cloud the sight, and, by the very knowledge of functions and processes, to bereave the student of the manly contemplation of the whole. The savant becomes unpoetic.

> *The Collected Works of Ralph Waldo Emerson*
> Volume I
> Nature
> Prospects (p. 39)

Fiske, John

...there are moments when one passionately feels that this cannot be all. On warm June mornings in green country lanes, with sweet pine-odours, wafted in the breeze which sighs through the branches, and cloud-shadows flitting over far-off blue mountains, while little birds sing their love-songs, and golden-haired children weave garlands of wild roses; or when in the solemn twilight we listen to wondrous harmonies of Beethoven and Chopin that stir the heart like voices from an unseen world; at such times one feels that the profoundest answer which science can give to our questionings is but a superficial answer after all.

> *The Unseen World, and Other Essays*
> The Unseen World
> Part II (p. 56)

Fort, Charles
Every science is a mutilated octopus. If its tentacles were not clipped to stumps, it would feel its way into disturbing contacts.

In Damon Knight
Charles Fort: Prophet of the Unexplained
A Charles Fort Sampler (p. vi)

Gill, Eric
Science is analytical, descriptive, informative. Man does not live by bread alone, but by science he attempts to do so. Hence the deadliness of all that is purely scientific.

Essays
Art, Section II (p. 13)

Havel, Václav
Modern science abolishes as mere fiction the innermost foundations of our natural world: it kills God and takes his place on the vacant throne so henceforth it would be science that would hold the order of being in its hand as its sole legitimate guardian and so be the legitimate arbiter of all relevant truth. People thought they could explain and conquer nature— yet the outcome is that they destroyed it and disinherited themselves from it.

In L. Wolpert
The Unnatural Nature of Science
Introduction (p. ix)

Heinlein, Robert A.
If it can't be expressed in figures, it is not science; it is opinion.

Time Enough for Love
Intermission (p. 257)

MacArthur, Robert H.
But not all naturalists want to do science; many take refuge in nature's complexity as a justification to oppose any search for patterns... Doing science is not such a barrier to feeling or such a dehumanizing influence as is often made out. It does not take the beauty from nature.

Geographical Ecology
Introduction (p. 1)

Raymo, Chet
...science is a spider's web. Confidence in any one strand of the web is maintained by the tension and resiliency of the entire web.

The Virgin and the Mousetrap
Chapter 16 (p. 144)

Schrödinger, Erwin
—who are we?... I consider this not only one of the tasks, but *the* task, of science, the only one that really counts.

Science and Humanism
The Alleged Break-Down of the Barrier between Subject and Object (p. 51)

Shapiro, Harry L.
Science, like organic life, has ramified by expanding into unoccupied areas and then adapting itself to the special requirements encountered there. And just as the diversified forms of animals, plants, and insects make evident by their morphology and their function the characteristics of ecological niches whose very existence might otherwise escape notice, so the diversity of techniques and concepts of scientific specialties by their very formulation reveal aspects of nature we would not have suspected. Anthropology, like other branches of science, has also embodied in its structure whole new worlds rich in insights into the development and nature of man.

American Anthropologist
Symposium on the History of Anthropology
The History and Development of Physical Anthropology (p. 371)
Volume 61, Number 3, 1959

Steward, J.H.
It is the unhappy lot of science that it must clear the ground of flimsy and fanciful structures built upon false premises and errors of fact before it can build anew.

Smithsonian Institution Annual Report
Petroglyphs of the United States
1936

Thomas, Lewis
The central task of science is to arrive, stage by stage, at a clearer comprehension of nature, but this does not mean, as it is sometimes claimed to mean, a search for mastery over nature.

Late Night Thoughts on Listening to Mahler's Ninth Symphony
Humanities and Science (p. 153)

Thoreau, Henry David
There is a chasm between knowledge and ignorance which the arches of science can never span.

The Writings of Henry David Thoreau
Volume I
A Week on the Concord and Merrimack Rivers
Sunday (p. 125)

Wolpert, Lewis
When we come to face the problems before us—poverty, pollution, overpopulation, illness—it is to science that we must turn, not to gurus. The arrogance of scientists is not nearly as dangerous as the arrogance that comes from ignorance.

New Scientist
In Mary Midgley
Can Science Save Its Soul? (p. 24)
Volume 135, Number 1832, 1 August 1992

SCIENTIFIC METHOD

Bauer, H.

One of the things wrong with the popular, classical definition of the scientific method is the implication that solitary people can successfully do good science, for example frame hypotheses and test them. In practice, however, the people who put forward the hypotheses are not usually the same people who apply the best test to them.

Scientific Literacy and the Myth of the Scientific Method
Chapter 3 (p. 52)

Russell, Bertrand

There are in science immense numbers of different methods, appropriate to different classes of problems; but over and above them all, there is something not easily definable, which may be called *the* method of science. It was formerly customary to identify this with the inductive method, and to associate it with the name of Bacon. But the true inductive method was not discovered by Bacon, and the true method of science is something which includes deduction as much as induction, logic and mathematics as much as botany and geology.

Mysticism and Logic
The Place of Science in a Liberal Education
Section II (p. 40)

Skinner, B.F.

Here was a first principle not formally recognized by scientific methodologists: When you run into something interesting, drop everything else and study it.

The American Psychologist
A Case History in Scientific Method (p. 223)
Volume 11, 1956

SCIENTIST

Feynman, Richard P.
The scientist has a lot of experience with ignorance and doubt and uncertainty, and this experience is of very great importance, I think...We have found...that in order to progress we must recognize our ignorance and leave room for doubt. Scientific knowledge is a body of statements of varying degrees of certainty—some most unsure, some nearly sure, but none *absolutely* certain.

What Do You Care What Other People Think? (p. 245)

Heinlein, Robert A.
Most "scientists" are bottle washers and button sorters.

Time Enough for Love
Intermission (p. 257)

Medawar, Peter
Scientists are people of very dissimilar temperaments doing different things in very different ways. Among scientists are collectors, classifiers and compulsive tidiers-up; many are detectives by temperament and many are explorers; some are artists and others artisans. There are poet-scientists and philosopher-scientists and even a few mystics.

Pluto's Republic (p. 116)

Menzel, Donald
Boyd, Lyle G.
The creative scientist, eternally curious, keeps an open mind toward strange phenomena and novel ideas, knowing that we have only begun to understand the universe we live in. He remembers, too, that Biot's discovery that meteorites were 'stones from the sky' was at first greeted with disbelief, and he hopes never to be guilty of similar obtuseness. But an open mind does not mean credulity or a suspension of the logical faculties that are man's most valuable asset.

The World of Flying Saucers (p. 289)

Taylor, A.M.

The three attributes of commitment, imagination, and tenacity seem to be the distinguishing marks of greatness in a scientist. A scientist must be as utterly committed to the pursuit of truth as the most dedicated of mystics; he must be as pertinacious in his struggle to advance into uncharted country as the most indomitable pioneers; his imagination must be as vivid and ingenious as a poet's or a painter's. Like other men, for success he needs ability and some luck; his imagination may be sterile if he has not a flair for asking the right questions, questions to which nature's reply is intelligible and significant.

Imagination and the Growth of Science
Chapter I (p. 5)

Weiss, Paul A.

Just like the painter, who steps periodically back from his canvas to gain perspective, so the laboratory scientist emerges above ground occasionally from the deep shaft of his specialized preoccupation to survey the cohesive, meaningful fabric developing from innumerable component tributary threads, spun underground much like his own. Only by such shuttling back and forth between the worm's eye view of detail and the bird's eye view of the total scenery of science can the scientist gain and retain a sense of perspective and proportions.

In Arthur Koestler and J.R. Smythies
Beyond Reductionism
The Living System (p. 3)

Wilson, Edward O.

Scientists live and die by their ability to depart from the tribe and go out into an unknown terrain and bring back, like a carcass newly speared, some new discovery or fact or theoretical insight and lay it in front of the tribe; and then they all gather and dance around it. Symposia are held in the National Academy of Sciences and prizes are given. There is fundamentally no difference from a paleothic camp site celebration.

In Edward Lueders
Writing Natural History
Dialogue 1 (p. 25)

SEEDS

Baker, Henry
A ripe seed falling to the earth is in the condition of the ovum of an animal getting loose from its ovary and dropping into the uterus, and, to go on with the analogy, the juices of the earth swell and extend the vessels of the seed as the juices of the uterus do those of the ovum, till the seminal leaves unfold and perform the office of a placenta to the infant included plant; which, imbibing suitable and sufficient moisture, gradually extends its parts, fixes its own root, shoots above the ground, and may be said to be born.

Philosophical Transactions
The Discovery of a Perfect Plant in Semine (p. 451)
Number 457, 1740

Each seed includes a Plant: that Plant, again,
Has other Seeds, which other Plants contain:
Those other Plants have All their Seeds, and Those
More Plants again, successively, inclose.
Thus ev'ry single Berry that we find,
Has, really, in itself whole Forests of its Kind.

Philosophical Transactions
The Discovery of a Perfect Plant in Semine (p. 451)
Number 457, 1740

de La Mare, Walter
The seeds I sowed—
For weeks unseen—
Have pushed up pygmy
Shoots of green;
So frail you'd think
The tiniest stone
Would never let

A Glimpse be shown.

Rhymes and Verses
Seeds

Ruskin, John
The reason for seeds is that flowers may be; not the reason of flowers that seeds may be.

The Queen of the Air
II, Section 60 (p. 174)

Tabb, John Bannister
Bearing a life unseen,
Thou lingerest between
A flower withdrawn,
And—what thou ne'er shalt see—
A blossom yet to be
When thou art gone.

The Poetry of Father Tabb
Nature—Miscellaneous
The Seed

SEXUALITY

von Linne, Carl
The organs of generation, which in the animal kingdom are by nature generally removed from sight, in the vegetable kingdom are exposed to the eyes of all, and that when their nuptials are celebrated, it is wonderful what delight they afford to the spectator by their most beautiful colors and delicious odors.

Amoenitates Academicae
Oeconomia naturae
Volume 2, 1752 (p. 16)

By what mechanisms are the sexuality of the worker naked mole rats suppressed, and how does the queen exert her supremacy? Research at London's Institute of Zoology by Chris Faulkes and others shows surprisingly that the main mechanism are not pheromonal (chemical) as we might immediately suppose. Mysteriously, it is the queenly presence, her behaviour, that keeps the rest so firmly switched off; which one of the British researchers has called the 'Thatcher effect'.

New Scientist
Volume 131, No 1780, 3 August 1991 (p. 43)

SIZE

Schmidt-Nielsen, Knut
What is the ultimate limit to the size of land animals? Unfortunately, we are unable to give an adequate answer, and we cannot study the question by building a bigger elephant.

Journal of Experimental Zoology
Scaling in Biology: The Consequence of Size (p. 291)
Volume 194, 1975

SPECIALIZATION

Heinlein, Robert A.
A human being should be able to change a diaper, plan an invasion, butcher a hog, conn a ship, design a building, write a sonnet, balance accounts, build a wall, set a bone, comfort the dying, take orders, give orders, cooperate, act alone, solve equations, analyze new problems, pitch manure, program a computer, cook a tasty meal, fight efficiently, die gallantly. Specialization is for insects.

Time Enough For Love
Intermission (pp. 265–6)

Weiner, Jonathan
Specialization has gotten out of hand. There are more branches in the tree of knowledge than there are in the tree of life. A petrologist studies rocks; a pedologist studies soils. The first one sieves the soil and throws away the rocks. The second one picks up the rocks and brushes off the soil. Out in the field, they bump into each other only like Laurel and Hardy, by accident, when they are both backing up.

The Next One Hundred Years
Chapter 10 (pp. 198–9)

SPECIES

Blumenbach, Johann Friedrich
What is *species*? We say that animals belong to one and the same species if they agree so well in form and constitution that those things in which they differ may have arisen from degeneration...Now we come to the real difficulty, which is to set forth the characters by which *in the natural world* we may distinguish mere varieties from genuine species.

The Anthropological Treatises of Johann Friedrich Blumenbach
Section II (p. 188)

Darwin, Charles
Widely ranging species, abounding in individuals, which have already triumphed over many competitors in their own widely extended homes will have the best chance of seizing on new places, when they spread into new countries.

The Origin of Species
Chapter XII (p. 182)

Falk, Donald
We consider species to be like a brick in the foundation of a building. You can probably lose one or two or a dozen bricks and still have a standing house. But by the time you've lost 20 per cent of species, you're going to destabilize the entire structure. That's the way ecosystems work.

Christian Science Monitor
26 May 1989

Lyell, Charles
...species are abstractions, not realities—are like genera. Individuals are the only realities. Nature neither makes nor breaks moulds—all is plastic, unfixed, transitional, progressive, or retrograde.

There is only one great resource to fall back upon, a reliance that all is for the best, trust in God, a belief that truth is the highest aim, that if it destroys some idols it is better that they should disappear, that the intelligent ruler of the universe has given us this great volume as a privilege, that its interpretation is elevating.

In Leonard G. Wilson (ed.)
Sir Charles Lyell's Scientific Journals on the Species Question
Journal II
July 10, 1856 (p. 121)

Mayr, Ernst
We had an international conference in Rome in 1981 on the mechanisms of speciation. It was attended by many of the leading botanists, zoologists, paleontologists, geneticists, cytologists and biologists. The one thing on which they all agreed was that we still have no idea what happens genetically during speciation. That's a damning statement, but it's the truth.

Omni Magazine
February, 1983 (p. 78)

Morton, Ron L.
Species come, species go;
Some real fast, some real slow...

Music of the Earth
Chapter 10 (p. 267)

Nietzsche, Friedrich
The species does *not* grow into perfection: the weak again and again get the upper hand of the strong,—their large number, and their *greater cunning* are the cause of it.

In Alexander Tille
The Works of Friedrich Nietzsche
Volume XI
The Twilight of the Idols
Roving Expeditions of an Inopportune Philosopher
Section 14 (p. 174)

Terborgh, John
Species are the units of evolution.

Diversity and the Tropical Rain Forest
Chapter 1 (p. 6)

STRUCTURE

Barry, Martin
It has been usual to regard organic structure as manifesting design, because it shews adaptation to the function to be performed. It has also been suggested, that function may be equally well considered as the result of structure. And, truly so it may. Yet perhaps we are not required to shew the claim of either to priority; but may consider both structure and function,—harmonising, as they always do,—as having been simultaneously contemplated in the same design.

Edinburgh New Philosophical Journal
On the Unity of Structure in the Animal Kingdom (p. 116)
Volume 22, 1836–37

Lamarck, Jean Baptiste Pierre Antoine
Naturalists have remarked that the structure of animals is always in perfect adaptation to their functions, and have inferred that the shape and condition of their parts have determined the use of them. Now this is a mistake: for it may be easily proved by observation that it is on the contrary the needs and uses of the parts which have caused the development of these same parts, which have given birth to them when they did not exist, and which consequently have given rise to the condition that we find in each animal.

Zoological Philosophy
Chapter VII (p. 113)

SURVIVAL

Arnold, Edwin
How lizard fed on ant, and snake on him,
And kite on both; and how the fish-hawk robbed
The fish-tiger of that which it had seized;
The shrike chasing the bulbul, which did chase
The jewelled butterflies; till everywhere
Each slew a slayer and in turn was slain,
Life living upon death.

Edwin Arnold's Poetical Works
Volume I
The Light of Asia
First Book (p. 21)

Darwin, Charles
What a trifling difference must often determine which shall survive, and which shall perish.

In Francis Darwin (ed.)
Life and Letters of Charles Darwin
Volume I
Darwin to Asa Gray
September 5, 1857

Spencer, Herbert
This survival of the fittest, which I have here sought to express in mechanical terms, is that which Mr Darwin has called "natural selection, or the preservation of favoured races in the struggle for life".

The Principles of Biology
Part III, Chapter 12, Section 165

SYMMETRY

Carroll, Lewis
You boil it in sawdust;
You salt it in glue;
You condense it with locusts in tape;
Still keeping one principle object in view—
To preserve its symmetrical shape.

The Hunting of the Snark
Fit the Fifth (p. 56)

Déscartes, Rene
Anyone who, upon looking down at his bare feet, doesn't laugh, has either
no sense of symmetry or no sense of humor.

In Abdus Salam
Journal of Molecular Evolution
The Role of Chirality in the Origin of Life (p. 105)
Volume 33, Number 2, August 1991

Kaku, Michio
Thompson, Jennifer
...nature, at the fundamental level, does not just prefer symmetry in a
physical theory, nature *demands* it.

Beyond Einstein
Chapter 6 (p. 108)

SYNONYMY

Davis, P.H.
Heywood, V.H.
Uncritical citation of synonyms may lead to a repetition of errors. The monographer should accept nothing on trust that he can confirm personally.

Principles of Angiosperm Taxonomy
Chapter 9 (p. 294)

Schenk, E.T.
McMasters, J.H.
Clarity and brevity are among the essential attributes of a good synonymy; clarity should not, however, be sacrificed for the sake of brevity.

Procedures in Taxonomy
Chapter VI (p. 17)

SYNTHESIS

Mayr, Ernst
We didn't sit down together and forge a synthesis. We all knew each other's writings; all spoke with each other. We all had the same goal, which was simply to understand fully the evolutionary process...By combining our knowledge, we managed to straighten out all the conflicts and disagreements so that finally a united picture of evolution emerged.

In Pamela Weintraub (ed.)
The Omni Interviews
Darwin Flights (p. 47)

What is still lacking is a critical analysis of the writings of the architects of the synthesis.

The Growth of Biological Thought: Diversity, Evolution, Inheritance
Chapter 12 (p. 568)

The term "evolutionary synthesis" was introduced by Julian Huxley in *Evolution: The Modern Synthesis* to designate the general acceptance of two conclusions: gradual evolution can be explained in terms of small genetic changes ("mutations") and recombination, and the ordering of this variation by natural selection; and the observed evolutionary phenomena, particularly macroevolutionary processes and speciation, can be explained in a manner that is consistent with the known genetic mechanisms.

In E. Mayr and W.B. Provine
The Evolutionary Synthesis (p. 1)

Vivilov, N.I.
We are now entering an epoch of differential ecological, physiological and genetic classification. It is an immense work. The ocean of knowledge is practically untouched by biologists. It requires the joint labors of many different specialists—physiologists, cytologists, geneticists, systematists, and biochemists. It requires international spirit, the cooperative work of

408

investigators throughout the whole world...it will bring us logically to the next step: integration and synthesis.

In J. Huxley
The New Systematics
The New Systematics of Cultivated Plants (p. 565)

SYSTEM

Agassiz, Louis

... without a thorough knowledge of the habits of animals, it will never be possible to ascertain with any degree of precision the true limits of all those species which descriptive zoologists have of late admitted with so much confidence into their works. After all, what does it matter to science, that thousands of species more or less should be described and entered in our systems, if we know nothing about them?

Essay on Classification
Chapter I
Section XVI (p. 66)

de Queiroz, K.
Donoghue, M.J.

If the goal of systematics is to depict relationships accurately, then any tradition that interferes with this goal should be abandoned.

Cladistics
Phylogenetic Systematics of Nelson's Version of Cladistics
Volume 4, Number 4, December 1988 (p. 332)

Hennig, W.

In order to be able to judge correctly the position of systematics in the field of biology and the role that it is called upon to play in the solution of the basic problems of this science, one must first make clear that there is a systematics not only in biology, but that it is rather an integrating part of any science whatever. It is surprising and peculiar to see to what degree the original significance of this concept has been forgotten in biology in the course of the fundamentally inadmissible but now general limitation of the concept of systematics to a particular subdivision of the science as a whole.

In George Gaylord Simpson
Principles of Animal Taxonomy
Systematics, Taxonomy, Classification, Nomenclature (p. 6)

Mayr, Ernst

The systematist who studies the factors of evolution wants to find out how species originate, how they are related, and what this relationship means. He studies species not only as they are, but also their origin and changes. He tries to find his answers by observing the variability of natural populations under different external conditions and he attempts to find out which factors promote and which inhibit evolution. He is helped in this endeavor by his knowledge of the habits and the ecology of the studied species.

Systematics and the Origin of Species
Chapter I (p. 11)

Novacek, M.J.

Thus a paleontologist unearthing skeletons in an Asian desert and a molecular biologist sequencing a strand of deoxyribonucleic acid (DNA) can both claim to be systematists if they share an interest in how species are related and how they arose over time. All these issues depend on theories of patterns of descent, or organisms branching off from each other in a way that accurately reflects their histories. When such theories continue to successfully explain new observations, they form the basis for many statements about the biological world.

In N. Eldredge
Systematics, Ecology and the Biodiversity Crisis
The Meaning of Systematics and the Biodiversity Crisis (p. 103)

Simpson, George Gaylord

Systematics is the scientific study of the kinds and diversity of organisms and of any and all relationships among them.

Principles of Animal Taxonomy
Systematics, Taxonomy, Classification, Nomenclature (p. 7)

Thompson, W.R.

The good systematist develops what the medieval philosophers called a *habitus*, which is more than a habit and is better designated by its other name of *secunda natura*. Perhaps, like a tennis player or a musician, he works best when he does not get too introspective about what he is doing.

Canadian Entomology
The Philosophical Foundation of Systematics (p. 5)
Volume 84, 1952

von Goethe, Johann Wolfgang

Natural system—a contradiction in terms. Nature has no system; she has, she *is* life and its progress from an unknown center toward an unknowable goal. Scientific research is therefore endless, whether one proceed analytically into minutiae or follow the trail as a whole, in all its breadth and height.

Botanical Writings
Problems (p. 116)

TAXONOMY

Cain, A.J.
It is not extraordinary that young taxonomists are trained like performing monkeys, almost wholly by imitation, and that in only the rarest cases are they given any instruction in taxonomic theory.

In George Gaylord Simpson
Principles of Animal Taxonomy
Preface (p. vii)

Constance, L.
Plant taxonomy has not outlived its usefulness: it is just getting under way on an attractively infinite task.

American Journal of Botany
Plant Taxonomy in an Age of Experiment (p. 92)
Volume 44, Number 1, January 1957

Heywood, V.H.
In these days when Molecular Biology is beginning to be seen as a restricted science, narrowing our vision by concentrating on the basic uniformity of organisms at the macromolecular level, the need for taxonomists to draw attention to the enormous diversity and variation of this earth's biota becomes more and more pressing.

In Tod. F. Stuessy
Plant Taxonomy
Plant Taxonomy (p. xvii)

Kevan, D.K. McE.
Bad taxonomy, of which there has been plenty, persists. Unlike bad chemistry or bad physiology, of which there has probably been equally as much, it cannot be ignored; it must be undone and redone. Poor taxonomy is not only an ill unto itself; it is contagious, often with a very long incubation period...One assumes that when [experimental biologists] state that they used 5 ml ethanol, they were not using 6 ml of methanol;

and yet, if the experimental animal is wrongly identified, what are the grounds for such an assumption?

Canadian Entomology
The Place of Classical Taxonomy in Modern Systematic Entomology (p. 1212)
Volume 105, 1973

Rollins, R.C.

In other words, the field of taxonomy in a way epitomizes the work of all other branches of biology centered on the organism itself, and brings the varied factual information from them to bear on the problems of interrelationship, classification and evolution. Thus taxonomy, as has been aptly remarked, is at once the alpha and omega of biology.

American Journal of Botany
Taxonomy of the Higher Plants (p. 188)
Volume 44, Number 1, January 1957

Schenk, E.T.
McMasters, J.H.

With the vast increase in numbers of known forms of animals and with the change in concepts of classification brought about by acceptance of the theory of evolution, the mechanics of modern taxonomy have become so complex as to discourage the beginning student.

Procedure in Taxonomy
Chapter I (p. 1)

Simpson, George Gaylord

Taxonomy is a science, but its application to classification involves a great deal of human contrivance and ingenuity, in short, of art. In this art there is leeway for personal taste, even foibles, but there are also canons that help to make some classifications better, more meaningful, more useful than others.

Principles of Animal Taxonomy
From Taxonomy to Classification (p. 107)

Stuessy, Tod F.

We as taxonomists celebrate diversity. We celebrate the wildness of the planet. We celebrate the numerous human attempts to understand this wilderness, and we mourn its loss through human miscalculation. We sense the aesthetic of life and much of our efforts are aimed at reflecting this composition. Above all we celebrate the challenges of being alive and dealing with the living world. There is no greater responsibility, privilege, nor satisfaction.

Plant Taxonomy
Epilogue (p. 406)

Wald, George
The most important thing about a name, after all, is that it remain attached to the thing it designates. One wishes that once a name had come into common use for an organism, it could be stabilized for the use of busy persons who want nothing but that each animal have a name.

In E.S. Guzman Barron (ed.)
Modern Trends in Physiology and Biochemistry
Biochemical Evolution (fn on p. 339)

TELEOLOGY

Ayala, Francisco J.

Biological evolution can however be explained without recourse to a Creator or a planning agent external to the organisms themselves. The evidence of the fossil record is against any directing force, external or immanent, leading the evolutionary process toward specified goals. Teleology in the stated sense is, then, appropriately rejected in biology as a category of explanation.

American Scientist
Biology as an Autonomous Science (p. 213)
Volume 56, Number 3, Autumn 1968

Henderson, Lawrence

Science has put the old teleology to death. Its disembodied spirit, freed from vitalism and all material ties, immortal, alone lives on, and from such a ghost science has nothing to fear.

The Fitness of the Environment
Chapter VIII (p. 311)

Reichenbach, Hans

Teleology is analogism, is pseudo explanation; it belongs in speculative philosophy, but has no place in scientific philosophy.

The Rise of Scientific Philosophy
Chapter 12 (p. 195)

von Bruecke

Teleology is a lady without whom no biologist can live. Yet he is ashamed to show himself with her in public.

In W.I.B. Beveridge
The Art of Scientific Investigation
Chapter V (p. 61)

TERMINOLOGY

Unknown
It has been long known
I haven't bothered to check the references.
It is immediately obvious that
Aren't I clever?
Possible therapeutic application
Please, please don't cut the funding now
It is known
I believe
It is believed
I think
It is generally believed
My colleagues and I think
There has been some discussion
Nobody agrees with me
It can be shown
Take my word for it
It is proven
It agrees with something mathematical
Of great theoretical importance
I find it interesting
Of great practical importance
This justifies my employment
Of great historical importance
This ought to make me famous
Some samples were chosen for study
The others didn't make sense
Typical results are shown
The best results are shown
Correct within order of magnitude
Wrong
The sample was put through two rounds of purification

417

After purification, the sample was dropped on the floor, slurped up with a pipette, and repurified.

The XYZ system was chosen as especially suited to show the predicted behavior

The guy in the next lab already had the system set up.

The values were obtained empirically

The values were obtained by accident

The results are inconclusive

The results seem to disprove my hypothesis

Additional work is required

Someone else can work out the details

It might be argued that

I have a good answer to this objection

The investigations proved rewarding

My grant has been renewed

Synthesised according to standard protocols

Purchased from Sigma

Thanks to Joe Blow for expert technical assistance and Jane Doe for valuable discussion

Thanks to Joe Blow for doing all the work and Jane Doe for telling me what it meant.

While it has not been possible to provide definite answers to these questions

The experiments didn't work out, but I figured I could at least get a publication out of it.

Mus musculus domesticus was chosen as especially suitable to test this hypothesis

Mus musculus domesticus is a lovely animal which is easy to study in the lab.

Accidentally strained during mounting

Dropped on the floor

Handled with extreme care

Not dropped on the floor throughout the experiments

Although some detail has been lost in reproduction, it is clear from the original micrograph

It is impossible to tell from the original micrograph

Presumably at longer times

I didn't take the time to find out.

The most reliable values are those of Jones

Jones was a student of mine.

It is generally believed that

A couple of other guys think so too.

It is clear that much additional work will be required before a complete understanding

I don't understand it

Unfortunately, a quantitative theory to account for these effects has not been formulated

Neither does anybody else
It is hoped that this work will stimulate further work in the field
This paper isn't very good, but neither are any of the others in this miserable subject
High purity
Composition unknown except for the exaggerated claims of the suppliers
A fiducial reference line on the specimen
A scratch.

Source Unknown

Look at this as a learning experience.
You're going to suffer.
Let me explain the format of the defense.
Let me make you even more nervous.
I'm here to lend you support.
I'm here to destroy you so you won't look smarter than me.
I found the overall concept interesting.
This is my token compliment before ripping your idea to shreds.
I would like to have had more time to study this.
I didn't read it.
I have some concerns about the theory upon which your study is based.
I hate the theory, but I can't insult the author so I'll insult your work instead.
There are some aspects of the study that I would like to hear more about.
I read it but I just don't remember anything about it.
Your hypotheses are not strongly enough linked to the existing literature.
You came up with an innovative idea and I want to make sure you never do it again.
Your research is an interesting extension of my own work.
Why didn't I think of this before you did?
You have failed to take into account some of the more relevant literature.
You failed to cite me.
I would like you to explain...
I don't know anything about this stuff so you'll have to explain it to me.
Your statistical results don't seem to support your hypothesis.
I don't understand statistics.
Your selection of statistical tests is rather simplistic.
I'm the only one here that understands statistics and I wanted to rub it in.
How did you ensure that you had drawn a random sample?
I had to come up with at least one question and this one always works.
This is a great topic for your thesis.
This is some grunge work that will help me get tenure.
You will be ready to write up soon, but need to do just one more experiment/program/chip.

You have now become a useful slave, and I am not about to let you graduate without doing more footwork for me.
Your funding is secure.
Maybe.
Your funding is probably OK.
Start worrying.
I'll see what we can do about funding.
Start looking for another advisor.
Think of this as an investment in skills that will be useful to you in your later career.
We're going to exploit you to the gills.
Don't listen to XYZ, just listen to me.
Both XYZ and I are fools, but I'm funding you.
Let's wrap this up.
I'm hungry.
Could you step out of the room while the committee comes to a decision?
We decided beforehand to give you your degree, but we still want to make you sweat some more.

Thesis Terminology
Source unknown

In a forest a fox bumps into a little rabbit, and says, "Hi, junior, what are you up to?"

"I'm writing a dissertation on how rabbits eat foxes," said the rabbit.

"Come now, friend rabbit, you know that's impossible!"

"Well, follow me and I'll show you."

They both go into the rabbit's dwelling and after a while the rabbit emerges with a satisfied expression on his face. Comes along a wolf. "Hello, what are we doing these days?"

"I'm writing the second chapter of my thesis, on how rabbits devour wolves."

"Are you crazy? Where is your academic honesty?"

"Come with me and I'll show you."

As before, the rabbit comes out with a satisfied look on his face and a diploma in his paw. Finally, the camera pans into the rabbit's cave and, as everybody should have guessed by now, we see a mean-looking, huge lion sitting next to some bloody and furry remnants of the wolf and the fox... The moral: It's not the contents of your thesis that are important—it's your PhD advisor that really counts.

Source unknown

TIME

Borland, Hal
Forget that second-ticking clock. Time is the seed
Waiting to fly from the milkweed pod. Time is the speed
Of a dragonfly. Time is the rabbit's desperate scut.
Time's dimensions are hidden in rocks,
In wind and rain, but never in clocks.

Borland Country (p. 5)

Carlyle, Thomas
That great mystery of TIME, were there no other; the illimitable, silent,
never-resting thing called Time, rolling, rushing on, swift, silent, like an
all-embracing ocean-tide, on which we and all the Universe swim like
exhalations, like apparitions which are, and then are not:...

Sartor Restarus & On Heroes
Hero Worship
Lecture I (p. 246)

Eliot, T.S.
Time present and time past
Are both perhaps present in time future,
And time future contained in time past.

Four Quartets
Burnt Norton
Stanza I

Hutton, J.
Time, which measures every thing in our idea, and is often deficient to our
schemes, is to nature endless and as nothing.

Transactions of the Royal Society of Edinburgh
Theory of the Earth
Volume 1, 1788 (p. 215)

Huxley, Thomas H.

Biology takes her time from geology. The only reason we have for believing in the slow rate of the change in living forms is the fact that they persist through a series of deposits which, geology informs us, have taken a long while to make. If the geological clock is wrong, all the naturalist will have to do is to modify his notions of the rapidity of change accordingly.

Quarterly Journal of the Geological Society London
Volume 25 (p. xxxviii)

Lamarck, Jean Baptiste Pierre Antoine

Time is insignificant and never a difficulty for Nature. It is always at her disposal and represents an unlimited power with which she accomplishes her greatest and smallest tasks.

Hydrogeology (p. 61)

Madách, Imre

LUCIFER: All things that live, endure for the same span;
    The century-old tree, and the one-day beetle,
    Grow conscious, joy and love, and pass away
    When they have reached their own appointed aims.
    Time does not move. 'Tis only we who change.
    A hundred years are but one brief day.

The Tragedy of Man
Scene I (p. 37)

Urey, H.

However, the evolution from inanimate systems of biochemical compounds, e.g., the proteins, carbohydrates, enzymes and many others, of the intricate systems of reactions characteristic of living organisms, and the truly remarkable ability of molecules to reproduce themselves seems to those most expert in the field to be almost impossible. Thus a time from the beginning of photosynthesis of two billion years may help to accept the hypothesis of the spontaneous generation of life.

Proceedings of the National Academy of Science
On the Early Chemical History of the Earth and the Origin of Life (p. 362)
Volume 38, 1952

Wald, George

Time is in fact the hero of the plot. The time with which we have to deal is of the order of two billion years. What we regard as impossible on the basis of human experience is meaningless here. Given so much time, the 'impossible' becomes possible, the possible probable, and the probable virtually certain. One has only to wait: time itself performs the miracles.

In A Scientific American Book
The Physics and Chemistry of Life
The Origin of Life (p. 12)

TREE

Borland, Hal
Only the unobservant sees nothing but trees in a forest. Any woodland is a complex community of plants and animal life with its own laws of growth and survival. But if you would know strength and majesty and patience, welcome the company of trees.

Beyond Your Doorstep
Chapter 4 (p. 75)

Cather, Willa
I like trees because they seem more resigned to the way they have to live than other things do.

O Pioneers!
Part II
Chapter VIII (p. 96)

Hay, John
They [trees] hang on from a past no theory can recover. They will survive us. The air makes their music. Otherwise they live in savage silence, though mites and nematodes and spiders teem at their roots, and though the energy with which they feed on the sun and are able to draw water sometimes hundreds of feet up their trunks and into their twigs and branches calls for a deafening volume of sound.

The Undiscovered Country
Living with Trees (p. 110)

Herbert, George
Great trees are good for nothing but shade.

Outlandish Proverbs

Isaiah 41:19
I will plant in the wilderness the cedar, the shittah tree, and the myrtle, and the oil tree; I will set in the desert the fir tree, and the pine, and the box tree together...

The Bible

Morris, George P.
Woodman, spare that tree!
Touch not a single bough!
In youth it sheltered me,
And I'll protect it now.

Poems
Woodman, Spare That Tree

Pownall, Thomas
The individual Trees of those Woods grow up, have their Youth, their old Age, and a Period to their Life, and die as we Men do. You will see many a Sapling growing up, many an old Tree tottering to its Fall, and many fallen and rotting away, while they are succeeded by others of their Kind, just as the Race of Man is: By this Succession of Vegetation this Wilderness is kept cloathed with Woods just as the human Species keeps the Earth peopled by its continuing Succession of Generations.

A Topographical Description of the Dominion of the United States
Section I
On the Face of the Country (p. 24)

Proust, Marcel
We have nothing to fear and a great deal to learn from trees, that vigorous and pacific tribe which without stint produces strengthening essences for us, soothing balms, and in whose gracious company we spend so many cool, silent and intimate hours.

Pleasures and Regrets
Regrets, Reveries, Changing Skies
Chapter XXVI (p. 165)

St Bernard
Believe me who have tried. Thou wilt find something more in woods than in books. Trees and rocks will teach what thou canst not hear from a master.

In A.C. Seward
Links with the Past in the Plant World
Chapter I (p. 1)

Shaw, George Bernard
Except during the nine months before he draws his first breath, no man manages his affairs as well as a tree does.

Man and Superman
Maxims for Revolutionists
The Unconscious Self

ACACIA

Browning, Elizabeth Barrett
A great acacia with its slender trunk
And overpoise of multitudinous leaves
(In which a hundred fields might spill their dew
And intense verdure, yet find room enough)
Stood reconciling all the place with green.

The Complete Poetical Works of Elizabeth Barrett Browning
Aurora Leigh
Book VI, l. 536–41

Dorr, Julia C.R.
Pluck the acacia's golden balls,
And mark where the red pomegranate falls.

Under the Palm-Trees

ALMOND

Preston, Margaret J.
White as the blossoms which the almond tree,
Above its bald and leafless branches bear.

The Royal Preacher
Stanza 5

APPLE

Hawthorne, Nathaniel
And what is more melancholy than the old apple-trees that linger about the spot where once stood a homestead, but where there is now only a ruined chimney rising out of a grassy and weed-grown cellar? They offer their fruit to every wayfarer—apples that are bitter-sweet with the moral of time's vicissitude.

Mosses from an Old Manse
The Old Manse (p. 15)

Thoreau, Henry David
It is remarkable how closely the history of the apple tree is connected with that of man.

The Writings of Henry David Thoreau
Volume V
Wild Apples (p. 290)

ASH

Lowell, Maria White
The ash her purple drops forgivingly
And sadly, breaking not the general hush;
The maple's swamps glow like a sunset sea,
Each leaf a ripple with its separate flush;
All round the wood's edge creeps the skirting blaze,
Of bushes low, as when, on cloudy days,
Ere the rain falls, the cautious farmer burns his brush.

An Indian Summer Reverie
Stanza 11

ASPENS

Unknown
At that awful hour of the Passion, when the Savior of the world felt deserted in His agony, when—

The sympathizing sun, his light withdrew, and wonder'd how the stars their dying Lord could view—

when earth, shaken with horror, rung the passing bell for Deity, and universal nature groaned; then from the loftiest tree to the lowliest flower all felt a sudden thrill, and trembling, bowed their heads, all save the proud and obdurate *aspen*, which said, 'Why should *we* weep and tremble? we trees, and plants, and flowers are pure and never sinned!'

Ere it ceased to speak, an involuntary trembling seized its every leaf, and the word went forth that it should never rest, but tremble on until the day of judgment.

Notes and Queries
Legend
First Series
Volume VI, Number 161

BANYAN

Bryant, Alice Franklin
Like a cathedral in some old world town
Rising above all mundane buildings, rears
The banyan tree, a growth of long slow years,
Towering above the palms. Its verdant crown
Fashions a far-spread roof, from which falls down
A diamond and tinted light with jeweled spears
Of sunbeam piercing through. The whole appears
An ornate Gothic pile of world renown.

Nature Magazine
The Banyan Tree (p. 265)
Volume 50, Number 5, May 1957

BAOBAB

de Saint-Exupéry, Antoine
Now there were some terrible seeds on the planet that was the home of the little prince; and these were the seeds of the baobab. The soil of that planet was infested with them. A baobab is something you will never, never be able to get rid of if you attend to it too late. It spreads over the

entire planet. It bores clear through it with its roots. And if the planet is too small, and the baobabs are too many, they split it to pieces...

The Little Prince
V (p. 21)

BEECH

Campbell, Thomas
Oh, leave this barren spot to me!
Spare, woodman, spare the beechen tree!

The Complete Poetical Works of Thomas Campbell
The Beech-Tree's Petition
Stanza I

BIRCH

Lowell, Maria White
Rippling through thy branches goes the sunshine,
Among thy leaves that palpitate forever,
And in the see, a pining nymph had prisoned
The soul, once of some tremulous inland river,
Quivering to tell her woe, but ah! dumb, dumb forever.

The Birch Tree

CHERRY

Longfellow, Henry Wadsworth
Sweet is the air with the budding haws, and the valley stretching for miles below
Is white with blossoming cherry-trees, as if just covered with lightest snow.

The Complete Writings of Henry Wadsworth Longfellow
Volume V
Christus
Golden Legend
Part IV (p. 265)

CHESTNUT

Ingelow, Jean
And when I see the chestnut letting
All her lovely blossoms falter down, I think

"Alas the day!"

Poems
The Warbling of Blackbirds

Lowell, Maria White
The chestnuts, lavish of their long-hid gold,
To the faint Summer, beggared now and old,
Pour back the sunshine hoarded 'neath her favoring eye.

An Indian-Summer Reverie
Stanza 10

CITRON

Milton, John
Awake, the morning shines, and the fresh field
Call us; we lose the prime, to mark how spring
Our tended Plants, how blows the Citron Grove,
What drops the Myrrhe, & what the balmie Reed,
How Nature paints her colours, how the Bee
Sits on the Bloom, extracting liquid sweet.

Paradise Lost
Book V, l. 20-25

COCONUT

Twain, Mark
I once heard a grouty Northern invalid say that a coconut tree might be poetical, possibly it was; but it looked like a feather-duster struck by lightning.

Roughing It
Volume 2, Chapter 23 (p. 215)

CYPRESS

Byron, George
Dark tree—still sad when others' grief is fled,
The only constant mourner o'er the dead!

The Complete Poetical Works
Volume III
The Giaour, l. 286

ELM

Longfellow, Henry Wadsworth
And the great elms o'erhead
Dark shadows wove on their aerial looms
Shot through with golden thread.

The Complete Writings of Henry Wadsworth Longfellow
Volume III
Hawthorne
Stanza 2

Tennyson, Alfred
In crystal vapour everywhere
Blue isles of heaven laugh'd between,
And far, in forest-deeps unseen,
The topmost elm-tree gather'd green
From draughts of balmy air.

The Complete Poetical Works of Tennyson
Sir Lancelot and Queen Guinevere
Stanza I

FIR

Milton, John
Kindles the gummy bark of Firr or Pine,
And sends a comfortable heat from farr,
Which might supply the Sun...

Paradise Lost
Book X, l. 1076–8

HEMLOCK

Longfellow, Henry Wadsworth
O hemlock-tree! O hemlock-tree! how faithful
    are thy branches!
Green not alone in summer time,
But in the winter's frost and rime!
O hemlock-tree! O hemlock-tree! how faithful
    are thy branches!

The Complete Writings of Henry Wadsworth Longfellow
Volume VI
The Hemlock Tree
Stanza 1

HOLLY

Burns, Robert
Green, slender, leaf-clad holly-boughs
Were twisted gracefu', round her brows;
I took her for some Scottish Muse,
By that same token;
And come to stop those reckless vows,
Would soon be broken.

The Complete Poetical Works of Robert Burns
The Vision
First Duan, Stanza 9

LARCH

Hemans, Felicia
I have looked on the hills of the stormy North,
And the larch has hung all his tassels forth...

The Poetical Works of Mrs Felicia Hemans
The Voice of Spring
Stanza 3

LINDEN

Heine, Heinrich
If thou lookest on the lime-leaf,
Thou a heart's form wilt discover;
Therefore are the lindens ever
Chosen seats of each fond lover.

Book of Songs
New Spring
Number 23, Stanza 3 (p. 110)

LOTUS

Hayne, Paul H.
Where drooping lotos-flowers, distilling balm,
Dream by the drowsy streamlets sleep hath crown'd,
While Care forgets to sigh, and Peace hath balsamed Pain.

Sonnet
Pent in this Common Sphere

Pope, Alexander
A spring there is, whose silver waters show
Clear as a glass the shining sands below:
A flowering lotos spreads its arms above,
Shades all the banks, and seems itself a grove.

Sappho to Phaon
l. 177

MAHOGANY

Thackeray, William Makepeace
Christmas is here;
Winds whistle shrill,
Icy and chill,
Little care we;
Little we fear
Weather without,
Sheltered about
The Mahogany-Tree.

The Complete Poems of W.M. Thackeray
The Mahogany-Tree

MAPLE

English, Thomas Dunn
That was a day of delight and of wonder,
While lying the shade of the maple-trees under—
He felt the soft breeze at its frolicsome play;
He smelled the sweet odor of newly mown hay...

The Select Poems of Dr Thomas Dunn English
Under the Trees

OAK

Dryden, John
The monarch oak, the patriarch of the trees,
Shoots rising up, and spreads by slow degrees.
Three centuries he grows, and three he stays
Supreme in state; and in three more decays.

The Poetical Works of John Dryden
Tales From Chaucer
Palamon and Arcite
Book III, l. 1058

PALM

Bailey, L.H.
The heavier palms are the big game of the plant world.

Gentes Herbarium
Palms, and their Characteristics
3, Fasc. 1

Longfellow, Henry Wadsworth
As the palm-tree standeth so straight and so tall,
The more the hail beats, and the more the rain falls.

The Complete Writings of Henry Wadsworth Longfellow
Volume VI
Annie of Tharaw
Translated from the German of Simon Dach, l. 11

Spenser, Edmund
First the high palme-tree, with braunches faire,
Out of the lowly vallies did arise,
And high shoote up their heads into the skyes.

The Complete Poetical Works of Edmund Spenser
Virgils Gnat
l. 190–2

PEAR

Ingelow, Jean
The great white pear-tree dropped with dew from leaves
And blossom, under heavens of happy blue.

Poems
Songs with Preludes
Wedlock

PINE

Heine, Heinrich
A pine tree standeth lonely
On a far norland height:
It slumbereth, while round it
The snow falls thick and white.

Book of Songs
Lyrical Interlude
Number 34 (pp. 63–4)

Lowell, Maria White
The pine is the mother of legends.

The Growth of a Legend

Taylor, Bayard
Ancient Pines,
Ye bear no record of the years of man.
Spring is your sole historian...

The Poetical Works of Bayard Taylor
The Pine Forest of Monterey
Stanza 4

POPLAR

Bulwer-Lytton, Edward
Trees that, like the poplar, lift upwards all their boughs, give no shade and no shelter, whatever their height. Trees the most lovingly shelter and shade us, when, like the willow, the higher soar their summits, the lowlier droop their boughs.

What Will He Do With It?
Volume II
Book XI, Chapter X
Introductory lines (p. 359)

REDWOOD

Steinbeck, John
The redwoods once seen, leave a mark or create a vision that stays with you always... It's not only their unbelievable stature, nor the color which seems to shift and vary under your eyes, no, they are not just like any trees we know, they are ambassadors from another time.

Travels with Charley
Part 3 (p. 169)

SLOE

Whitman, Sarah Helen
In the hedge the frosted berries glow,
The scarlet holly and the purple sloe.

Poems
A Day of the Indian Summer

SPICE

Sterling, John
The Spice Tree lives in the garden green,
Beside it the fountain flows;
And a fair Bird sits the boughs between,
And sings his melodious woes.

Poems
The Spice Tree
Stanza 1

SYCAMORE

Ingelow, Jean
You night-moths that hover where honey brims over
From sycamore blossoms...

Songs of Seven
Seven Times Three
Stanza 3

THORN

Burns, Robert
Beneath the milk-white thorn that scents the ev'ning gale.

The Complete Poetical Works of Robert Burns
The Cotter's Saturday Night
Stanza IX

TULIP-TREE

Bryant, William Cullen
The tulip-tree, high up,
Opened, in airs of June, her multitude
Of golden chalices to humming-birds
And silken-winged insects of the sky.

Poems
The Fountain
Stanza 3

WILLOW

Thackeray, William Makepeace
Know ye the willow-tree,
Whose grey leaves quiver,
Whispering gloomily
To yon pale river?

The Complete Poems of W.M. Thackeray
The Willow-Tree

YEW

Wordsworth, William
Of vast circumference and gloom profound
This solitary Tree! A living thing
Produced too slowly ever to decay;
Of form and aspect too magnificent
To be destroyed.

The Complete Poetical Works of William Wordsworth
Yew-Trees

TRUTH

Darwin, Charles

The truth will not penetrate a preoccupied mind.

In Francis Darwin (ed.)
More Letters of Charles Darwin
Volume I
Darwin to Hooker
July 28, 1868 (p. 305)

Thomas, Lewis

The only solid piece of scientific truth about which I feel totally confident
is that we are profoundly ignorant about nature.

The Medusa and the Snail
The Hazard of Science (p. 73)

Wilson, Edward O.

... if history and science have taught us anything, it is that passion and
desire are not the same as truth. The human mind evolved to believe
in the gods. It did not evolve to believe in biology. Acceptance of the
supernatural conveyed a great advantage throughout prehistory, when
the brain was evolving. Thus it is in sharp contrast to biology, which was
developed as a product of the modern age and is not underwritten by
genetic algorithms. The uncomfortable truth is that the two beliefs are not
factually compatible. As a result those who hunger for both intellectual
and religious truth will never acquire both in full measure.

Consilience: The Unity of Knowledge
Chapter 11 (p. 262)

VARIATION

Darwin, Charles
...the number of intermediate varieties, which must have formerly existed, [must] be truly enormous. Why then is not every geological formation and every stratum full of such intermediate links? Geology assuredly does not reveal any such finely graduated organic chain; and this, perhaps, is the most obvious and gravest objection which can be urged against the theory.

The Origin of Species
Chapter X (p. 152)

Pallister, William
What shall we say of a plot of ground
Planted in similar seed,
Where thousands of similar plants are found
But one is a new type indeed;
When dissimilar comes from similar,
And freedom has its hour,
When the scion is not as ancestors are,
What is this latent power?

Poems of Science
De Ipsa Natura
Variation (p. 213)

Peirce, C.S.
The endless variety in the world has not been created by law. It is not the nature of uniformity to originate variation, nor of law to beget circumstance.

Collected Papers
Volume VI
Chapter 6, Section 2 (p. 373)

Waddington, C.H.

To suppose that the evolution of the wonderfully adapted biological mechanisms has depended only on a selection out of a haphazard set of variations, each produced by blind chance, is like suggesting that if we went on throwing bricks together into heaps, we should eventually be able to choose ourselves the most desirable house.

The Listener
13 February 1952

VIVISECTION

Shaw, George Bernard
Animals dislike being vivisected, but they also dislike being forced to bear burdens and draw loads. The difference is not in the pain endured by the animals, but in the fact that whereas there is no doubt than an intelligent horse would consent to do a reasonable quantity of work for its living if it were capable of economic reasoning, just as men do, it is equally certain that no horse would on any terms submit to vivisection. On this ground the vivisector violates the moral law.

In Brian Tyson
Bernard Shaw's Book Reviews
Volume I
Two Novels of Modern Society (p. 28)

WATER

Coleridge, Samuel T.
Water, water, everywhere,
And all the boards did shrink;
Water, water, everywhere,
Nor any drop to drink.

In Max J. Herzberg (ed.)
Narrative Poems
The Ancient Mariner
Part II, Stanza 9

Dalyell, J. Graham
On descending from terrestrial objects to the inhabitants of the waters, infinitely new and interesting matter is presented for the contemplative physiologist. Myriads of beings, alike singular in structure and properties, appear in their peculiar element, all actuated by the resistless impulse of nature; avoiding danger, seeking subsistence, rendering the weaker a prey.

Observations on Planariae

de Saint-Exupéry, Antoine
Water, thou hast no taste, no color, no odor; canst not be defined, art relished while ever mysterious. Not necessary to life, but rather life itself, thou fillest us with a gratification that exceeds the delight of the senses.

Wind, Sand and Stars
Prisoner of the Sand (p. 184)

Eiseley, Loren
If there is magic on this planet, it is contained in water.

The Immense Journey
The Flow of the River (p. 15)

Herbert, Sir Alan
The rain is plentious but, by God's decree,

Only a third is meant for you and me;
Two-thirds are taken by the growing things
Or vanish Heavenward on vapour's wings:
Nor does it mathematically fall
With social equity on one and all.
The population's habit is to grow
In every region where the water's low:
Nature is blamed for failings that are Man's,
And well-run rivers have to change their plans.

Source unknown

Norse, Elliot A.
In every glass of water we drink, some of the water has already passed through fishes, trees, bacteria, worms in the soil, and many other organisms, including people... Living systems cleanse water and make it fit, among other things, for human consumption.

In R.J. Hoage (ed.)
Animal Extinctions
The Value of Animal and Plant Species for
Agriculture, Medicine, and Industry (p. 62)

CARE FOR A GLASS OF WATER?
I PREPARED IT MYSELF..!

Strauss, Maurice B.

In the beginning the abundance of the sea
Led to profligacy.
The ascent through the brackish waters of the estuary
To the salt-poor lakes and ponds
Made immense demands
Upon the glands.
Salt must be saved, water is free.
In the never-ending struggle for security,
Man's chiefest enemy.
According to the bard of Stratford on the Avon,
The banks were climbed and life established on dry land
Making the incredible demand
Upon another gland
That water, too, be saved.

Body Water in Man
Salt and Water
Chapter XII (p. 238)

van Helmont, Joan-Baptista

That all plants immediately and substantially stem from the element water alone I have learnt from the following experiment. I took an earthen vessel in which I placed two hundred pounds of earth dried in an oven, and watered with rain water. I planted in it the stem of a willow tree weighing five pounds. Five years later it had developed a tree weighing one hundred and sixty-nine pounds and three ounces. Nothing but rain (or distilled water) had been added. The large vessel was placed in earth and covered by an iron lid with a tin-surface that was pierced with many holes. I have not weighed the leaves that came off in the four autumn seasons. Finally I dried the earth in the vessel again and found the same two hundred pounds of it diminished by about two ounces. Hence one hundred and sixty-four pounds of wood, bark and roots had come up from water alone.

In William H. Brock
The Norton History of Chemistry
Introduction (p. xxi)

Walton, Izaak

And an ingenious Spaniard says, that rivers and the inhabitants of the watery element were made for wise men to contemplate, and fools to pass by without consideration... for you may note, that the waters are Nature's storehouse, in which she locks up her wonders.

In Marston Bates
The Natural History of Mosquoitoes
Chapter VIII (p. 112)

WETLANDS

Beebe, William
The marsh, to him who enters it in a receptive mood, holds, besides mosquitoes and stagnation,—melody, the mystery of unknown waters, and the sweetness of Nature undisturbed by man.

Log of the Sun
Night Music of the Swamp (p. 172)

Lanier, Sidney
Ye marshes, how candid and simple and nothing-withholding and free
Ye publish yourselves to the sky and offer yourselves to the sea!

The Marshes of Glynn

WILDERNESS

Berry, Wendell
There does exist a possibility that we can live more or less in harmony with our native wilderness; I am betting my life that such a harmony is possible. But I do not believe that it can be achieved simply or easily or that it can ever be perfect, and I am certain that it can never be made, once and for all, but it is the forever unfinished lifework of our species.

Home Economics
Preserving Wilderness (pp. 138-9)

Hopkins, Gerard Manley
What would the world be, once bereft
Of wet and of wilderness? Let them be left,
O let them be left, wilderness and wet;
Long live the weeds and the wilderness yet.

In W.H. Gardner and N.H. MacKenzie (eds)
The Poems of Gerard Manley Hopkins
Inversnaid
Stanza 4

Lindbergh, Charles A.
In wilderness I sense the miracle of life, and behind it our scientific accomplishments fade to trivia.

Life
The Wisdom of Wilderness (p. 10)
Volume 63, Number 25, December 22, 1967

Thoreau, Henry David
In wildness is the preservation of the world.

The Atlantic Monthly
Walking (p. 665)
Volume 9, Number 56, June 1862

WILDLIFE

Borland, Hal

The newcomer to the country will find the first signs of "wild life" in his own house. Even before he explores the dooryard he can sharpen his eyes indoors. He may be surprised at the outsiders who want to share that house with him.

Beyond Your Doorstep
Chapter 1 (p. 1)

Hornaday, William T.

And yet the game of North America does not belong wholly and exclusively to the men who kill! The other ninety-seven per cent of the People have vested rights in it... Posterity has claims upon it that no man can ignore... A continent without wild life is like a forest with no leaves on the trees.

Our Vanishing Wild Life
Preface (p. ix)

Myers, Norman

Without knowing it, we utilize hundreds of products each day that owe their origin to wild animals and plants. Indeed our welfare is intimately tied up with the welfare of wildlife. Well may conservationists proclaim that by saving the lives of wild species, we may be saving our own.

A Wealth of Wild Species
Wild Species (p. 3)

Prince Philip

Miners used to take a canary around the coal mines to warn them when the air was so foul that the canary died. This is the importance of wildlife to us; because if wildlife dies it is our turn next. If any part of the life of this planet is threatened, all is threatened. If you say "not interested" to wildlife conservation then you are signing your own death warrant.

The Times (London)
May 17, 1988

WORMS

Beebe, William
There are many ways of considering a flatworm. A Creator might rightly be quoted, "He saw that it was good." To an ant accidentally blundering into its slime, the worm would be a certain, evil death. A bird would give it no second glance for its flesh is worse than inedible.

High Jungle
Chapter X (p. 171)

Blake, William
O rose, thou art sick!
The invisible worm
That flies in the night,
In the howling storm,
Has found out thy bed
Of crimson joy,
And his dark secret love
Does thy life destroy.

The Complete Poetry and Prose of William Blake
Songs of Experience
The Sick Rose

Boone, J. Allen
...if you should ever encounter me walking along a dirt road and should see me pause, lift my hat and bow to the direction of the ground, you will know that I am paying my respects to a passing earthworm.

Kinship With All Life
Wormy Ways (p. 123)

Darwin, Charles
Worms have played a more important part in the history of the world than most persons would at first suppose.

The Formation of Vegetable Mould, Through the Action of Worms,
With Observations of Their Habits

Eaton, Burnham
The earthworm who, described as lowly,
Grinds, like the gods, exceedingly slowly,
Doth also grind exceedingly small.
By diligent, continual
And through subterranean toil,
He doth homogenize the soil.

Nature Magazine
H-O-M-G-E-N-I-Z-A-T-I-O-N (p. 41)
Volume 50, Number 1, January 1957

Garstang, Walter
The Onchosphere or Hexacanth was not designed for frolic,
His part may be described perhaps as coldly diabolic:
He's born amid some gruesome things, but this should count for virtue,
That steadily, 'gainst fearful odds, he plies his task—to hurt you.
. . .
He's now a *Cysticerus* in the muscles of a pig,
With just a sporting chance of getting to grow up big.
If you'll consent to eat your pork half-raw or underdone,
His troubles will be over, and a Tapeworm will have won:
He'll cast his anchors out, and on your best digested food
Will thrive, and bud and endless chain to raise a countless brood.

Larval Forms
The Onchosphere
Stanzas 1, 4 (p. 37)

Gavenda, Walt
Wormy apples at the grocery,
Used to make consumers panic,
Now they sell at twice the price,
'Cause wormy apples are organic.

Source unknown

Isaiah 51:8
The moth shall eat them up like a garment, and the worm shall eat them
like wool.

The Bible

Martinson, Harry Edmund
Who really respects the earthworm,
the farm worker far under the grass in the soil.

In Robert Bly
Friends, You Drank Some Darkness
The Earthworm (p. 139)

Pallister, William
Then the WORMS seven thousand of species can show,
All segmented, possessing a system of nerves:
Life becoming more conscious, beginning to know;
The small earthworm is soils great economy serves,
Bringing earth to the surface, returning again,
Even thus, he has buried old cities for men!

Poems of Science
Beginnings
Animal Life (p. 139)

Pliny
Nature crying out and speaking to country people in these words: Clown, wherefore dost thou behold the heavens? Why dost thou seek after the stars? When thou art now werry with short sleep, the nights are troublesome to thee. So I scatter little stars in the grass, and I shew them in the evening when thy labour is ended, and thou art miraculously allured to look upon them when thous passest by: Dost thou not see how a light like fire is covered when she closeth her wings, and she carrieth both night and day with her.

In Thomas Moffett
The Theater of Insects
Glow-Worms

The Taylor Family
No little worm, you need not slip
Into your hole, with such a skip;
Drawing the gravel as you glide
On to your smooth and slimy side.

Original Poems for Infant Minds
The Worm

Unknown
A seventh grade Biology teacher arranged a demonstration for his class. He took two earth worms and in front of the class he did the following: He dropped the first worm into a beaker of water where it dropped to the bottom and wriggled about. He dropped the second worm into a beaker

of Ethyl alcohol and it immediately shriveled up and died. He asked the class if anyone knew what this demonstration was intended to show them.

A boy in the second row immediately shot his arm up and, when called on said: "You're showing us that if you drink alcohol, you won't have worms."

Source Unknown

I wish I were a glow-worm
A glow-worm's never glum,
How can you be unhappy
When a light shines out your bum.

Source unknown

Did you know that all animals went to Noah's Ark in pairs?
Yes, except the worms. They came in apples.

Source unknown

ZOO

Esar, Evan
[Zoo] Another place where people may visit but animals are barred.

Esar's Comic Dictionary

Hediger, Heini
Zoo biology is still a very young science and today many zoos are still run without the faintest idea that it exists. In some places no thought is given as to what the present role of a zoo either is or should be.

Man and Animal in the Zoo
Chapter 2 (p. 55)

One of the most frequent misconceptions which is constantly met in the zoo is the business of regarding the animals as prisoners. This is as false and old-fashioned as if in these days everybody still thought that radio and television sets contained little men who talked, sang and danced inside the sets.

Man and Animal in the Zoo
Chapter 3 (p. 99)

Wynne, Annette
Excuse us, Animals in the Zoo,
I'm sure we're very rude to you;
Into your private house we stare
And never ask you if you care;
And never ask you if you mind.
Perhaps we really are not kind:
I think it must be hard to stay
And have folks looking in all day,
I wouldn't like my house that way.

All Through the Year
Excuse Us, Animals in the Zoo

ZOOLOGY

Bierce, Ambrose
HIPPOGRIFF, *n*. An animal (now extinct) which was half horse and half griffin. The griffin was a compound creature, half lion and half eagle. The hippogriff was, therefore, only one quarter eagle, which is $2.50 in gold. Zoology is full of surprises.

The Devil's Dictionary

ZOOLOGY, *n*. The science and history of the animal kingdom, including its king, the House Fly ("*Musca maledicta*"). The father of Zoology was Aristotle, as is universally conceded, but the name of its mother has not come down to us.

The Devil's Dictionary

Bock, W.J.
Communication—information exchange—among zoologists is the core of zoological nomenclature; everything else pales in the light of the importance of communication.

Bulletin of the American Museum of Natural History
History and Nomenclature of Avian Family Group Names (p. 8)
Volume 221, 1994

Elton, Charles
...the discoveries of Darwin, himself a magnificent field naturalist, had the remarkable effect of sending the whole zoological world flocking indoors, where they remained hard at work for fifty years or more, and whence they are now beginning to put forth cautious heads again into the open air.

Animal Ecology
Chapter I (p. 3)

Esar, Evan
[Zoologist] The only one who can tell the difference between a white zebra with black stripes and a black zebra with white stripes.

Esar's Comic Dictionary

Feynman, Richard P.
I began to read the paper. It kept talking about extensors and flexors, the gastrocnemius muscle, and so on. This and that muscle were named, but I had not the foggiest idea of where they were located in relation to the nerves or to the cat. So I went to the librarian in the zoology section and asked her if she could find me a map of the cat.

"A *map* of the *cat*, sir?" she asked horrified. "You mean a *zoological chart*!"

From then on there were rumors about a dumb biology student who was looking for "a map of the cat".

Surely You're Joking, Mr. Feynman!
A Map of a Cat? (p. 72)

Kavalevski, V.O.
And so, the task of modern zoology consists in this; it should acquaint us with the entire variety of animal forms which populate our world, not in terms of a disorganized multitude from which this or that form happens to catch our attention, but as a structured whole, in which each form occupies a designated place, so one can instantly note and critically analyze all the particularities of each separate member; it should show us the inner structure of these groups and of their individual members, and in what relationship they stand to members of other groups; it should present the history of each member, beginning with its [first] appearance;... it should open the ancient tombs of the earth and demonstrate to us the endless series of ancestors and relatives which proceed those animals which we now see.

In William Coleman and Camille Limoges (eds)
Studies in History of Biology
Volume 2
Kovalevskii and Paleontology (pp. 112–13)

Polanyi, Michael
The existence of animals was not discovered by zoologists, nor that of plants by botanists, and the scientific value of zoology and botany is but an extension of man's pre-scientific interests in animals and plants.

Personal Knowledge
Chapter 6 (p. 139)

Queneau, Raymond

In the dog days while I was in a bird cage at feeding time I noticed a young puppy with a neck like a giraffe who, like the toad, ugly and venomous, wore yet a precious beaver upon his head. This queer fish obviously had a bee in his bonnet and was quite bats, he started yak-yakking at a wolf in sheep's clothing claiming that he was treading on his dogs with his beetle-crushers, but the sucker got a flea in his ear; that foxed him, and quiet as a mouse he ran like a hare for a perch.

I saw him again later in front of the Zoo with a young buck who was telling him to bear in mind a certain drill about his fevers.

Exercises in Style
Zoological (p. 179)

BIBLIOGRAPHY

Abingdon, Alexander. *Bigger & Better Boners*. The Viking Press, New York. 1952.

Ackerman, Diane. *The Moon by Whale Light*. Random House, New York. 1991.

Adams, Abby. *The Gardner's Gripe Book*. Workman Publishing, New York. 1995.

Adams, Douglas. *Dirk Gentley's Holistic Detective Agency*. Simon and Schuster, New York. 1987.

Adams, Henry. *The Education of Henry Adams*. The Heritage Press, New York. 1942.

Agassiz, Louis. *Essay on Classification*. The Belknap Press of Harvard University Press, Cambridge, MA. 1962.

Agassiz, Louis. *Geological Sketches*. Houghton, Mifflin and Company, Boston. 1886.

Agassiz, Louis and Gould, A.A. *Principles of Zoology*. Gould and Lincoln, Boston. 1859.

Aldrich, Thomas Bailey. *The Poems of Thomas Bailey Aldrich*. Houghton, Mifflin and Company, Boston. 1885.

Aldrich, Thomas Bailey. *The Queen of Sheba*. James R. Osgood and Company, Boston. 1877.

Allen, Ethan. *Reason the Only Oracle of Man*. Scholars' Facsimiles & Reprints, New York. 1940.

Altmann, Jeanne. *Baboon Mothers and Infants*. Harvard University Press, Cambridge. 1980.

Arber, Agnes. *The Mind and the Eye*. Cambridge University Press, Cambridge. 1954.

Ardrey, Robert. *African Genesis*. Atheneum, New York. 1986.

Ardrey, Robert. *The Social Contract*. Atheneum, New York. 1970.

Aristotle. 'Physics' in *Great Books of the Western World*. Volume 8. Encyclopaedia Britannica, Inc., Chicago. 1952.

Aristotle. 'On the Heavens' in *Great Books of the Western World*. Volume 8. Encyclopaedia Britannica, Inc., Chicago. 1952.

Aristotle. 'On the Parts of Animals' in *Great Books of the Western World*. Volume 9. Encyclopaedia Britannica, Inc., Chicago. 1952.

Armstrong, Martin D. 'Two Italian Gardens' in *The Atlantic Monthly*. Volume CX. September 1912.

Arnold, Edwin. *Edwin Arnold's Poetical Works*. Volume I. Roberts Brothers, Boston. 1889.

Arnold, Matthew. *The Poetical Works of Matthew Arnold*. Thomas Y. Crowell & Co., New York. 1897.

Asimov, Isaac. *A Choice of Catastrophes*. Hutchinson, London. 1979.

Atherton, Gertrude. *Senator North*. John Land: The Bodley Head, New York. 1900.

Atkinson, Brooks. *Once Around the Sun*. Harcourt, Brace and Company, New York. 1951.

Austin, Mary. *The Children Sing in the Far West*. Houghton, Mifflin Company, Boston. 1928.

Awiakta, Marilou. *Selu: Seeking the Corn-Mother's Wisdom*. Fulcrum Publishing, Golden. 1993.

Ayala, Francisco J. 'Biology as an Autonomous Science' in *American Scientist*. Volume 56, Number 3. Autumn 1968.

Bacon, Francis. 'Novum Organum' in *Great Books of the Western World*. Volume 30. Encyclopaedia Britannica, Inc., Chicago. 1952.

Bacon, Francis. *Of Gardens*. Swann Press. 1928.

Bailey, Philip James. *Festus: A Poem*. George Routledge and Sons, Limited, London. 1893.

Baker, Henry. 'The Discovery of a Perfect Plant in Semine' in *Philosophical Transactions*. Number 457. 1740.

Baker, Henry. *The Microscope Made Easy*. London. 1743.

Baker, Keith Michael. *Condorcet: From Natural Philosophy to Social Mathematics*. The University of Chicago Press, Chicago. 1975.

Balfour, James. *The Foundations of Belief*. Longmans, Green, and Co., London. 1912.

Barbellion, W.N.P. *The Journal of A Disappointed Man*. George H. Doran Company, New York. 1919.

Barron, E.S. Guzman. *Modern Trends in Physiology and Biochemistry*. Academic Press Inc., Publishers, New York. 1952.

Barry, Martin. 'On the Unity of Structure in the Animal Kingdomi' in *Edinburgh New Philosophical Journal*. Volume 22. 1836–7.

Bartram, William. *Travels and Other Writings*. The Library of America, New York. 1996.

Bates, Marston. *The Forest and the Sea*. Random House, New York. 1960.

Bates, Marston. *The Natural History of Mosquitoes*. Harper & Row, Publishers, New York. 1949.

Bauer, Henry H. *Scientific Literacy and the Myth of the Scientific Method*. University of Illinois Press, Urbana. 1992.

Baum, Harold. *The Biochemists' Handbook*. Pergamon Press, Oxford. 1982.

Bayliss, William Maddock. *Principles of General Physiology*. Longmans, Green, and Co., London. 1920.

Beard, Eva. 'Thomas Jefferson, Statesman and Scientist' in *Nature Magazine*. April 1958.

Beaumont, William. *William Beaumont: A Pioneer American Physiologist*. The C.V. Mosby Company, St Louis. 1981.

Beebe, William. *High Jungle*. Duell, Sloan and Pearce, New York. 1949.

Beebe, William. *The Log of the Sun*. Henry Holt and Company, New York. 1906.

Beecher, Henry Ward. *Life Thoughts*. Sheldon and Company, New York. 1869.

Beecher, Henry Ward. *Star Papers: Experiments of Art and Nature*. J.C. Derby, New York. 1855.

Beier, Ulli. *Yoruba Poetry*. Cambridge University Press, Cambridge. 1970.

Beilock, Richard P. *Beasts, Ballads, and Bouldingisms*. Transaction Books, New Brunswick. 1980.

Bell, E.T. *Men of Mathematics*. Simon and Schuster, New York. 1937.

Belloc, Hilaire. *Complete Verse*. Gerald Duckworth, London. 1970.

Benchley, Robert. *20,000 Leagues Under the Sea or David Copperfield*. Blue Ribbon Books, Garden City. 1946.

Benton, Allen H. and Werner, William E., Jr. *Field Biology and Ecology*. McGraw-Hill Book Company, New York. 1966.

Benzer, Seymour. *The Harvey Lectures*. Series 56. Academic Press, New York. 1961.

Berendzen, Richard. *Life Beyond Earth & the Mind of Man*. National Aeronautics and Space Administration, Washington, D.C. 1973.

Bergson, Henri. *Creative Evolution*. Henry Holt and Company, New York. 1911.

Bernal, J.D. *The Origin of Life*. The World Publishing Company, Cleveland. 1967.

Bernard, Claude. *An Introduction to the Study of Experimental Medicine*. Henry Schuman, Inc. 1949.

Berrill, N.J. *You and the Universe*. Fawcett Publications, Greenwich. 1958.

Berry, R.J., Crawford, T.J. and Hewitt, G.M. *Genes in Ecology*. Blackwell Scientific Publications, Oxford. 1992.

Berry, Wendell. *Home Economics*. North Point Press, San Francisco. 1987.

Berryman, John. *77 Dream Songs*. Farrar, Straus and Company, New York. 1964.

Beston, Henry. *The Outermost House*. Rinehart & Co., Inc., New York. 1949.

Beveridge, W.I.B. *The Art of Scientific Investigation*. William Heinemann Ltd, Melbourne. 1950.

Bierce, Ambrose. *The Devil's Dictionary*. Dover Publications, Inc., New York. 1958.

Bierce, Ambrose. *The Eyes of the Panther*. Jonathan Cape, London. 1928.

Billroth, Theodor. *The Medical Sciences in the German Universities*. The Macmillan Company, New York. 1924.

Bingham, Roger. *A Passion to Know: 20 Profiles in Science*. Charles Scribner's Sons, New York. 1984.

Blackie, John Stuart. *Musa Burschicosa*. Edmondtron and Douglas, Edinburgh. 1869.

Blackwood, Oswald. *Introductory College Physics*. John Wiley & Sons, Inc., New York. 1943.

Blake, William. *The Complete Poetry and Prose of William Blake*. University of California Press, Berkeley. 1982.

Blake, William. *The Letters of William Blake*. Rupert Hart-Davis, London. 1956.

Blanshard, Brand. *The Nature of Thought*. Volume 1. George Allen & Unwin Ltd, London. 1939.

Bloomfield, Robert. *The Farmer's Boy*. Houghton, Mifflin and Company, Boston. 1871.

Blumenbach, Johann Friedrich. *The Anthropological Treatises of Johann Friedrich Blumenbach*. Longman, Green, Longman, Roberts, & Green, London. 1865.

Bly, Robert. *Friends, You Drank Some Darkness*. Beacon Press, Boston. 1975.

Bly, Robert. *The Winged Life*. Sierra Club Books, San Francisco. 1986.

Bock, W.J. 'History and Nomenclature of Avian Family Group Names' in *Bulletin of the American Museum of Natural History*. Volume 221, 1994.

Bohr, Neils. 'Light and Life' in *Nature*. Volume 131, Number 3309. April 1, 1933.

Boone, J. Allen. *Kinship with All Life*. Harper & Brothers, Publishers, New York. 1954.

Booth, Mark. *What I Believe*. The Crossroad Publishing Company, New York. 1984.

Borges, Jorge Luis. *Other Inquisitions*. Simon and Schuster, New York. 1965.

Borkowski, L. *Selected Works*. North-Holland Publishing Company, Amsterdam. 1970.

Borkowski, L. *Studies in Logic*. North-Holland Publishing Company, Amsterdam. 1970.

Borland, Hal. *Beyond Your Doorstep*. Alfred A. Knopf, New York. 1965.

Borland, Hal. *Book of Days*. Alfred A. Knopf, New York. 1976.

Borland, Hal. *Borland Country*. J.B. Lippincott Company, Philadelphia. 1971.

Borland, Hal. *Sundial of the Seasons*. J.B. Lippincott, Company, Philadelphia. 1964.

Borlase, William. *The Natural History of Cornwall, The Air, Climate, Waters, Lakes, Seas and Tides.* Printed by W. Jackson for the author, London. 1758.

Born, Max. *Natural Philosophy of Cause and Chance.* Dover Publications, Inc., New York. 1964.

Born, Max. *The Born–Einstein Letters.* Translated by Irene Born. Walker and Company, New York. 1971.

Bosewell, James. *Life of Johnson.* Volume I. Oxford University Press, New York. 1933.

Botkin, D.B., Caswell, M.F., Estes, J.E. and Orio, A.A. *Changing the Global Environment: Perspectives on Human Involvement.* Academic Press, Boston. 1989.

Boulding, Kenneth E. *The Image.* The University of Michigan Press, Ann Arbor. 1956.

Boulton, James T. *The Selected Letters of D.H. Lawrence.* University Press, Cambridge. 1997.

Boyle, Robert. *The Sceptical Chymist.* J.M. Dent & Sons Ltd, London. 1944.

Bradbury, S. *The Microscope Past and Present.* Pergamon Press, Oxford. 1968.

Bradford, Gamaliel. *Darwin.* Houghton Mifflin Company, Boston. 1926.

Bradley, Mary Hastings. *On the Gorilla Trail.* D. Appleton and Company, New York. 1923.

Bramwell, Anna. *Ecology in the 20th Century: A History.* Yale University Press, New Haven. 1989.

Bridgman, Helen Bartlett. *Gems.* Brooklyn. 1916.

Bridgeman, P.W. 'The New Vision of Science' in *Harpers Magazine.* Volume 158. March 1929.

Bridges, Robert. *Poetical Works of Robert Bridges.* Volume II. Smith, Elders & Co., London. 1899.

Brillouin, Leon. *Scientific Uncertainty and Information.* Academic Press, New York. 1964.

Brock, William H. *The Norton History of Chemistry.* W.W. Norton & Company, New York. 1992.

Bronowski, J. *The Common Sense of Science.* William Heinemann Ltd, Melbourne. 1951.

Bronowski, J. *The Identity of Man.* The Natural History Press, Garden City. 1972.

Brooks, Paul. *The House of Life: Rachel Carson at Work.* Houghton Mifflin Company, Boston. 1972.

Brooks, W.K. 'Heredity and Variation: Logical and Biological' in *Proceedings of the American Philosophical Society.* Volume 45. April 20, 1906.

Brophy, Brigid. *Don't Never Forget.* Holt, Rinehart and Winston, New York. 1966.

Brower, David. *For Earth's Sake*. Peregrine Smith Books, Salt Lake City. 1990.

Browning, Elizabeth Barret. *The Complete Poetical Works of Elizabeth Barret Browning*. Houghton Mifflin Co., Boston. 1900.

Browning, Robert. *The Poems and Plays of Robert Browning*. The Modern Library, New York. 1934.

Bryant, Alice Franklin. 'The Banyan Tree' in *Nature Magazine*. Volume 50, Number 5. May 1957.

Bryant, William Cullen. *Poems*. D. Appleton & Co., New York. 1874.

Buber, Martin. *At the Turning Point*. Farrar, Strauss and Young, New York. 1952.

Buffon, Comte de Georges, Louis Leclerc. *Natural History, General and Particular*. Volume I. Printed for W. Strahan and T. Cadell, London. 1785.

Buffon, Comte de Georges, Louis Leclerc. *Natural History, General and Particular*. Volume VI. Printed for W. Strahan and T. Cadell, London. 1785.

Buhl, David. 'Chemical Constituents of Interstellar Clouds' in *Nature*. Volume 234. December 1971.

Bullard, Fred M. *Volcanoes of the Earth*. University of Texas Press, Austin. 1984.

Bulloch, W. 'Obituary Notice of Deceased Members' in *Journal of Pathology and Bacteriology*. Volume 40, Number 3. May 1935.

Bulwer-Lytton, Edward. *What Will He Do With It?* George Routledge and Sons, London. 1880.

Burns, Robert. *The Complete Poetical Works of Robert Burns*. Houghton Mifflin Company, Boston. 1897.

Burroughs, John. 'In the Noon of Science' in *The Atlantic Monthly*. Volume CX. September 1912.

Burroughs, John. 'Expression' in *The Atlantic Monthly*. Volume VI, Number XXXVII. November 1860.

Burroughs, John. *Harvest of a Quiet Eye*. Tamarack Press, Madison. 1976.

Burroughs, John. *Signs and Seasons*. Houghton, Mifflin and Company, Boston. 1886.

Burroughs, John. *Songs of Nature*. Doubleday, Page & Co., Garden City. 1912.

Burroughs, John. *Time and Change*. Houghton Mifflin Company, Boston. 1912.

Burroughs, John. *Ways of Nature*. Books for Libraries Press, Freeport. 1971.

Butler, Samuel. *Hudibras*. Clarendon Press, Oxford. 1967.

Byron, George. *The Complete Poetical Works*. Volume III. Clarendon Press, Oxford. 1981.

Byron, George. *The Complete Poetical Works*. Volume V. Clarendon Press, Oxford. 1981.

Callahan, Daniel. *The Tyranny of Survival*. Macmillan Publishing Co., Inc., New York. 1973.

Calvin, M. and Gazenko, O.G. *Foundations of Space Biology and Medicine*. Volume I. National Aeronautics and Space Administration. Washington, D.C. 1975.

Campbell, Thomas. *The Complete Poetical Works of Thomas Campbell*. Crosby, Nichols, Lee & Company, Boston. 1860.

Camus, Albert. *Notebooks: 1935–1951*. Marlowe & Company, New York. 1998.

Camus, Albert. *The Myth of Sisyphus*. Translated by Justin O'Brien. Penguin Books, Middlesex. 1984.

Canetti, Elias. *The Human Province*. Translated by Joachim Neugroschel. The Seabury Press, New York. 1978.

Capra, Fritjof. *The Turning Point*. Simon and Schuster, New York. 1982.

Carlton, J.T. 'Nonextinction of Marine Invertebrates' in *American Zoologist*. Volume 33. 1993.

Carlyle, Thomas. *Latter-Day Pamphlets*. Scribner, Welford and Company, New York. 1872.

Carlyle, Thomas. *Past and Present*. Charles Scribner's Sons, New York. 1843.

Carlyle, Thomas. *Sartor Resartus*. Frederick A. Stokes Company, New York. 1893.

Carlyle, Thomas. *Sartor Resartus & On Heroes*. J.M. Dent & Sons Ltd, London. 1929.

Carr, Archie. *The Windward Road*. Alfred A. Knopf, New York. 1956.

Carroll, Lewis. *The Complete Works of Lewis Carroll*. The Modern Library, New York. 1936.

Carroll, Lewis. *The Hunting of the Snark*. Macmillan and Co., New York. 1891.

Carson, Rachel. *Silent Spring*. Houghton, Mifflin Company, Boston. 1962.

Carson, Rachel. *The Edge of the Sea*. Houghton Mifflin Company, Boston. 1955.

Carson, Rachel. *The Sea Around Us*. Oxford University Press, New York. 1961.

Cather, Willa. *O Pioneers!*. Bantam Books, New York. 1989.

Chapman, Frank M. *Bird-Life*. D. Appleton and Company, New York. 1903.

Chapman, George T.L., and Tweddle, Marilyn N. *William Turner: A New Herball*. Part I. Cambridge University Press, Cambridge. 1989.

Chargaff, Edwin. *Essays on Nucleic Acids*. Elsevier Publishing Company, Amsterdam. 1963.

Chaucer, Geoffrey. 'The Canterbury Tales' in *Great Books of the Western World*. Volume 22. Encyclopaedia Britannica, Inc., Chicago. 1952.

Chaucer, Geoffrey. 'The Parliament of Fowls' in *The Complete Works of Geoffrey Chaucer*. Oxford University Press, London. nd.

Chesterton, G.K. *Orthodoxy*. John Lane Company, New York. 1918.

Chesterton, G.K. *The Defendant*. J.M. Dent & Sons Ltd, London. 1932.

Child, L. Maria. *Letters from New York*. C.S. Francis and Company, New York. 1844.

Chiras, Daniel D. *Lessons from Nature*. Island Press, Washington, D.C. 1992.

Churchill, Charles. *The Poems of Charles Churchill*. Volume II. The King's Printers. 1933.

Clare, John. *The Rural Muse*. The Mid Northumberland Arts Group and Carcanet New Press, Ashington. 1982.

Clements, Frederic E. and Shelford, Victor E. *Bio-Ecology*. John Wiley & Sons, Inc., New York. 1939.

Cleveland, John. *The Poems of John Cleveland*. The Grafton Press, New York. 1903.

Close, Frank, Michael Marten and Christine Sutton. *The Particle Explosion*. Oxford University Press, New York. 1987.

Cloud, Preston. *Oasis in Space*. W.W. Norton & Company, New York. 1988.

Cohen, H. Floris. *The Scientific Revolution*. The University of Chicago Press, Chicago. 1994.

Cohen, Joel. 'Mathematics as Metaphor' in *Science*. Volume 172. 14 May 1971.

Coleman, William and Limoges, Camille. *Studies in History of Biology*. Volume 2. The Johns Hopkins University Press, Baltimore. 1978.

Coleridge, Mary E. *Gathered Leaves*. Constable and Company, London. 1910.

Coleridge, Samuel Taylor. 'Table Talk, Part II' in *The Collected Works of Samuel Taylor Coleridge*. Volume 14. Princeton University Press, Princeton. 1990.

Coleridge, Samuel Taylor. *The Complete Poetical Works of Samuel Taylor Coleridge*. Volume I. Clarendon Press, Oxford. 1912.

Collard, Patrick. *The Development of Microbiology*. Cambridge University Press, Cambridge. 1976.

Collingwood, R.G. *The Idea of History*. Clarendon Press, Oxford. 1946.

Collins, Wilkie. *The Moonstone*. International Collectors Library, Garden City. 1900.

Colum, Padriac. *Poems*. The Macmillan Company, New York. 1932.

Commoner, Barry. *The Closing Circle: Nature, Man & Technology*. Bantam Books, New York. 1972.

Compton, Karl Taylor. *A Scientist Speaks*. Undergraduate Association, Massachusetts Institute of Technology, Cambridge, MA. 1955.

Comte, Auguste. *The Positive Philosophy of Auguste Comte*. Translated by Harriet Martineau. Trübner and Co., London. 1875.

Conklin, E.G. 'A Generation's Progress in the Study of Evolution' in *Science*. Volume 80, Number 2068. August 17, 1934.

Connolly, Cyril. *The Unquiet Grave*. Hamish Hamilton, London. 1951.

Constance, L. 'Plant Taxonomy in an Age of Experiment' in *American Journal of Botany*. Volume 44, Number 1. January 1957.

Cook, Eliza. *The Poetical Works of Eliza Cook*. Sorin & Ball, Philadelphia. 1848.

Cornford, F.M. 'Innumerable Worlds in Presocratic Philosophy' in *The Classical Quarterly*. January 1934.

Cornwall, Barry. *The Poetical Works of Milman, Bowles, Wilson, and Barry Cornwall*. A. and W. Galignani, Paris. 1829.

Cotton, E.H. *Has Science Discovered God?* Thomas Y. Crowell Company, New York. 1931.

Cowdry, Edmund V. *General Cytology*. The University of Chicago Press, Chicago. 1924.

Cowper, William. *The Poetical Works of William Cowper*. John W. Lovell Company, New York. nd.

Crabbe, George. *Poems*. Volume I. Cambridge University Press, Cambridge. 1905.

Cramer, F. *Chaos and Order*. Translated by D.I. Loewus. VCH, Weinheim. 1993.

Cranch, Christopher Pearse. *Collected Poems of Christopher Pearse Cranch*. Scholars' Facsimiles & Reprints, Gainsville. 1971.

Crick, Francis. *Life Itself*. Simon and Schuster, New York. 1981.

Crick, Francis. *Of Molecules and Men*. University of Washington Press, Seattle. 1966.

Crick, Francis. *What Mad Pursuit*. Basic Books, Inc., Publishers, New York. 1988.

Crichton, M. *The Andromeda Strain*. Alfred A. Knopf, New York. 1969.

Croll, Oswald. *Basilica Chymica*. Impensis Godefirdi Tampachii, Francofurti. 1608.

Croly, George. *The Poetical Works of the Rev. George Croly*. H. Colburn and R. Bently, London. 1830.

Crosbie, John S. *Crosbie's Dictionary of Puns*. Harmony Books, New York. 1977.

Crothers, Samuel McChord. *The Gentle Reader*. Books for Libraries Press, Freeport. 1972.

Crow, J.F. 'Genetic Effects of Radiation' in *Bulletin of the Atomic Scientists*. Volume XIV, Number 1. January 1958.

Crowson, R.A. *The Biology of the Coleoptera*. Academic Press, London. 1981.

Cudmore, L.L. Larison. *The Center of Life*. The New York Times Book Co., New York. 1977.

Cuppy, Will. *How to Become Extinct*. Dover Publications, Inc., New York. 1941.

Cuppy, Will. *How to Get From January to December*. Henry Holt and Company, New York. 1951.

Cussler, Clive and Dirgo, Craig. *The Sea Hunters*. Simon & Schuster, New York. 1996.

da Vinci, Leonardo. *Leonardo da Vinci's Notebooks*. Translated by Edward McCurdy. Empire State Book Co., New York. 1935.

da Vinci, Leonardo. *A Treatise on Painting*. George Bell & Sons, London. 1901.

Danforth, C.H. 'Genetics and Anthropology' in *Science*. Volume 79. 1934.

Darwin, Charles. 'The Descent of Man' in *The Origin of Species and The Descent of Man*. The Modern Library, New York. 1936.

Darwin, Charles. 'The Origin of Species' in *Great Books of the Western World*. Volume 49. Encyclopaedia Britannica, Inc., Chicago. 1952.

Darwin, Charles. *The Variation of Animal and Plants Under Domestication*. D. Appleton & Co., New York. 1883.

Darwin, Charles. *The Voyage of the Beagle*. P.F. Collier & Son Corporation, New York. 1937.

Darwin, Charles. 'The Various Contrivances by Which Orchids are Fertilized by Insects' in *The Works of Charles Darwin*. Volume 17. New York University Press, New York. 1988.

Darwin, Erasmus. *The Botanic Garden*. Jones & Company, London. 1825.

Darwin, Francis. *More Letters of Charles Darwin*. Volume I. J. Murray, London. 1903.

Darwin, Francis. *More Letters of Charles Darwin*. Volume II. J. Murray, London. 1903.

Darwin, Francis. *The Life and Letters of Charles Darwin*. Volume I. D. Appleton and Company, New York. 1896.

Darwin, Francis. *The Life and Letters of Charles Darwin*. Volume II. D. Appleton and Company, New York. 1896.

Davis, P.H. and Heywood, V.H. *Principles of Angiosperm Taxonomy*. D. Van Nostrand Company, Inc., Princeton. 1963.

Dawkins, Richard. *The Blind Watchmaker*. W.W. Norton & Company, New York. 1986.

Dawkins, Richard. 'The Necessity of Darwinism' in *New Scientist*. Volume 94, Number 1301. 15 April 1982.

Day, Clarence Jr. *This Simian World*. Alfred A. Knopf, New York. 1922.

Day, Richard Edwin. *Poems*. Cassell & Company, Limited, New York. 1888.

Dayton, P.K., Mordida, B.J., and Bacon, F. 'Polar Marine Communities' in *American Zoologist*. Volume 34, 1994.

de Beer, G.R. *Evolution: Essays on Aspects of Evolutionary Biology Presented to Professor E.S. Goodrich on His Seventieth Birthday*. Clarendon Press, Oxford. 1938.

de Chardin, Teilhard. *The Phenomenon of Man*. Harper & Row, Publishers, New York. 1965.

de Duve, Christian. *Vital Dust*. Basic Books, New York. 1995.

de Fontenelle, Bernard. *Conversations on the Plurality of Worlds*. Printed by J.C. Undee, London. 1803.

de La Mare, Walter. *Rhymes and Verses*. Holt, Rinehart and Winston, New York. 1947.

de Queiroz, Kevin and Donoghue, Michael J. 'Phylogenetic Systematics and the Species Problem' in *Cladestics*. Volume 4, Number 4. December 1988.

de Saint-Exupéry, Antoine. *The Little Prince*. Harcourt, Brace and Company, New York. 1943.

de Saint-Exupéry, Antoine. *Wind, Sand and Stars*. Harcourt, Brace & World, Inc., New York. 1967.

Debus, Allen G. *The French Paracelsians*. Cambridge University Press, Cambridge. 1991.

Delbrück, Max. 'A Physicist Looks at Biology' in *Transactions of the Connecticut Academy of Sciences*. Volume 38. 1949.

Dennett, Daniel C. *Consciousness Explained*. Little, Brown and Company, Boston. 1991.

Depew, D.J. and Wever, B.H. *Evolution at a Crossroads: The New Biology and the New Philosophy of Science*. MIT Press, Cambridge. 1985.

Derry, Gregory N. *What Science Is and How It Works*. Princeton University Press, Princeton. 1999.

Dewey, John. *The Influence of Darwin on Philosophy*. Indiana University Press, Bloomington. 1965.

Dickens, Charles. *Nicholas Nickleby*. T.B. Peterson & Brothers, Philadelphia. 1865.

Dickens, Charles. *Pickwick Papers*. Oxford University Press, London. 1948.

Dickinson, Emily. *Poems by Emily Dickinson*. Little, Brown and Company, Boston. 1948.

Dickinson, G. Lowes. *The Meaning of Good*. J.M. Dent & Co., London. nd.

Dillard, Annie. *Pilgrim at Tinker Creek*. Harper's Magazine Press, New York. 1974.

Diolé, Philippe. *The Errant Ark*. Translated by J.F. Bernard. G.P. Putnam's Sons, New York. 1974.

Disraeli, Benjamin. *Lothair*. Longmans, Green, and Co., London. 1970.

Disraeli, Benjamin. *Tancred*. Volume I. Henry Colburn, Publisher, London. 1847.

Doane, Rennie W. *Insects and Disease*. Henry Holt and Company, New York. 1910.

Dobzhansky, Theodosius. 'On Methods of Evolutionary Biology and Anthropology' in *American Scientist*. Volume 45. 1957.

Dobzhansky, Theodosius. 'Biology, Molecular and Organismic' in *American Zoologist*. Volume 4. 1964.

Dobzhansky, Theodosius. 'Changing Man' in *Science*. Volume 155, Number 3761. January 1967.

Dobzhansky, Theodosius. 'Nothing in Biology Makes Sense Except in the Light of Evolution' in *The American Biology Teacher*. Volume 35, Number 3. March 1973.

Dobzhansky, Theodosius. *Genetics of the Evolutionary Process*. Columbia University Press, New York. 1970.

Dobzhansky, Theodosius. *Genetics and the Origin of Species*. Columbia University Press, New York. 1951.

Dobzhansky, Theodosius. *The Biology of Ultimate Concern*. The New American Library, New York. 1967.

Donne, John. *Complete English Poems*. St Martin's Press, New York. 1971.

Douglas, Mary and Wildavsky, Aaron. *Risk and Culture*. University of California Press, Berkeley. 1982.

Doyle, Arthur Conan. *The Complete Sherlock Holmes*. Doubleday & Company, Inc., Garden City. 1960.

Drake, Daniel. *An Introductory Lecture on the Means of Promoting the Intellectual Improvement of the Students and Physicians, of the Valley of the Mississippi*. Published by the Class, Louisville. 1844.

Drexler, K.E. *Engines of Creation: the Coming Era of Nanotechnology*. Anchor Press, Garden City. 1986.

Driesch, Hans. *The History & Theory of Vitalism*. Macmillan and Co., Limited, London. 1914.

Dryden, John. *The Poetical Works of John Dryden*. Volume II. James Nichol, Edinburgh. 1855.

du Noüy, Lecomte. *Human Destiny*. The New American Library, New York. 1960.

Dubos, René. *Louis Pasteur*. Charles Scribner's Sons, New York. 1976.

Dubos, René. *The Dreams of Reason*. Columbia University Press, New York. 1961.

Dukas, Helen and Hoffman, Banesh. *Albert Einstein: The Human Side*. Princeton University Press, Princeton. 1979.

Dunn, L.C. *Genetics in the 20th Century*. The Macmillan Company, New York. 1960.

Dunn, R.A. and Davisdon, R.A. 'Pattern Recognition in Biological Classification' in *Pattern Recognition*. Volume 1. 1968.

Durant, Will and Durant, Ariel. *The Lessons of History*. Simon and Schuster, New York. 1968.

Durrell, Gerald M. *The Overloaded Ark*. The Viking Press, New York. 1953.

Eaton, Burnham. 'H-O-M-O-G-E-N-I-Z-A-T-I-O-N' in *Nature Magazine*. Volume 50, Number 1. January 1957.

Eckermann, J.P. *Conversations with Goethe*. Frederick Ungar Publishing Co., New York. 1964.

Eckert, Allan W. *Wild Season*. Little, Brown and Company, Boston. 1967.

Eddington, A.S. *The Nature of the Physical World*. The Macmillan Company, New York. 1948.

Eddington, A.S. *The Philosophy of Physical Science*. The Macmillan Company, New York. 1939.

Edwards, Llewellyn Nathaniel. *A Record of History and Evolution of Early American Bridges*. University Press, Orono. 1959.

Edwards, R.Y. 'Research: A Museum Cornerstone' in *Occasional Papers of the British Columbia Provisional Museum*. Volume 25. 1985.

Egner, Robert E. and Denonn, Lester E. *The Basic Writings of Bertrand Russell*. Simon and Schuster, New York. 1961.

Ehlers, Vernon. 'Human Cloning' in *Congressional Record-House*. Volume 143, No. 26. 4 March 1997.

Ehrenreich, Barbara. *The Worst Years of Our Lives*. Pantheon Books, New York. 1990.

Ehrlich, Gretel. *The Solace of Open Spaces*. Viking, New York. 1985.

Einstein, Albert. *Cosmic Religion*. Covici Friede, Publishers, New York. 1931.

Eiseley, Loren. *The Immense Journey*. Vintage Books, New York. 1957.

Eiseley, Loren. *The Star Thrower*. Times Books, New York. 1978.

Eldredge, Niles. *Systematics, Ecology, and the Biodiversity Crisis*. Columbia University Press, New York. 1992.

Eliot, George. *Scenes of Clerical Life*. Clarendon Press, Oxford. 1985.

Eliot, George. 'Daniel Deronda' in *The Writings of George Eliot*. Volume 16. Houghton Mifflin Company, Boston. 1908.

Eliot, T.S. *Four Quartets*. Harcourt, Brace and Company, New York. 1943.

Eliot, T.S. *Old Possum's Book of Practical Cats*. Harcourt Brace Javonvich Publishers, New York. 1982.

Elliott, Ebenezer. *The Poetical Works of Ebenezer Elliott*. William Tait, Edinburgh. 1840.

Ellis, Havelock. *The Dance of Life*. Houghton Mifflin Company, Boston. 1923.

Elton, Charles. *Animal Ecology*. The Macmillan Company, New York. 1936.

Embury, Emma C. *The Poems of Emma C. Embury*. Hurd and Houghton, New York. 1869.

Emerson, Ralph Waldo. *Collected Poems and Translations*. The Library of America, New York. 1994.

Emerson, Ralph Waldo. *Essays. First and Second Series*. Houghton Mifflin Company, Boston. 1883.

Emerson, Ralph Waldo. *Society and Solitude*. Houghton Mifflin, Boston. 1898.

Emerson, Ralph Waldo. *The Conduct of Life*. Ticknor and Fields, Boston. 1860.

Emerson, Ralph Waldo. *The Collected Works of Ralph Waldo Emerson*. Volume I. Harvard University Press, Cambridge. 1971.

Emmeche, Claus. *The Garden in the Machine*. Translated by Steven Sampson. Princeton University Press, Princeton. 1994.

English, Thomas Dunn. *The Select Poems of Dr. Thomas Dunn English*. Published by Private Subscription, Newark. 1894.

Erskine, John. *The Moral Obligation to be Intelligent*. Duffield and Company, New York. 1916.

Esar, Evan. *20,000 Quips & Quotes*. Barnes & Noble Books, New York. 1995.

Esar, Evan. *Esar's Comic Dictionary*. Doubleday & Company, Inc., Garden City. 1983.

Evans, Howard Ensign. *Pioneer Naturalist: The Discovery and Naming of North American Plants and Animals*. Henry Holt and Company, New York. 1993.

Evans, Howard Ensign. *The Pleasures of Entomology*. Smithsonian Institution Press, Washington, D.C. 1985.

Fairchild, H.N. *Religious Trends in English Poetry*. Volume I. Columbia University Press, New York. 1939.

Fauset, Jessie. *The Chinaberry Tree*. Negro University Press, New York. 1931.

Fermi, Laura. *Atoms in the Family*. The University of Chicago Press, Chicago. 1954.

Ferris, G.F. 'The Principles of Systematic Entomology' in *Stanford University Publications: Biological Science*. Volume 5. 1928.

Feynman, Richard. *Surely You're Joking, Mr. Feynman!*. W.W. Norton & Company, New York. 1985.

Feynman, Richard. *The Character of Physical Law*. British Broadcasting Company, London. 1965.

Field, Eugene. *The Complete Tribune Primer*. Mutual Book Company, Boston. 1901.

Fishback, Margaret. *I Take It Back*. E.P. Dutton & Co., Inc., New York. 1935.

Fischer, Ernst Peter. *Beauty and the Beast*. Plenum Trade, New York. 1999.

Fisher, R.A. *The Genetical Theory of Natural Selection*. Clarendon Press, Oxford. 1930.

Fiske, John. *The Destiny of Man*. Houghton, Mifflin and Company, Boston. 1884.

Fiske, John. *The Unseen World, and Other Essays*. Houghton, Mifflin and Company, Boston. 1876.

Flack, Jerry. 'Quotations in the Classroom' in *Teaching K-8*. Volume 24, Number 3. November/December 1993.

Flammarion, Camille. *Popular Astronomy*. Chatto & Windus, London. 1894.

Flanders, Michael and Minale, Marcello. *Creatures Great and Small...* Holt, Rinehart and Winston, New York. 1964.

Flannery, Maura C. 'Biology is Beautiful' in *Perspectives in Biology and Medicine*. Volume 35, Number 3. Spring 1992.

Flaubert, Gustave. *Dictionary of Accepted Ideas*. Max Reinhardt, London. 1954.

Fletcher, Joseph. *The Ethics of Genetic Control*. Anchor Books, Garden City. 1974.

Florian, Douglas. *Insectlopedia*. Harcourt Brace & Company, San Diego. 1998.

Florio, John. *His First Fruites*. Da Capo Press, Amsterdam. 1969.

Forbes, Edward. *The Natural History of the European Seas*. John Van Voorst, London. 1859.

Foreman, Dave. *Confessions of an Eco-Warrior*. Harmony Books, New York. 1991.

Fosdick, W.W. *Ariel and Other Poems*. Bunce & Brothers, Publishers, New York. 1855.

Freud, Sigmund. *Beyond the Pleasure Principle*. Boni and Liveright Publishers, New York. 1924.

Frisch, Karl von. *Bees: Their Vision, Chemical Sense, and Language*. Cornell University Press, Ithaca. 1950.

Frost, Robert. *The Poetry of Robert Frost*. Holt, Rinehart and Winston, New York. 1967.

Fuller, R. Buckminster. *Nine Chains to the Moon*. Southern Illinois University Press, Carbondale. 1963.

Futuyma, Douglas J. *Evolutionary Biology*. Sinauer Associates, Inc., Sunderland. 1979.

Gahan, A.B. 'The Role of the Taxonomist in Present Day Entomology' in *Entomological Scoiety of Washington Proceedings*. Volume 25. 1923.

Gallagher, Winifred. *The Power of Place*. Poseidon Press, New York. 1993.

Gardner, John. *A Child's Bestiary*. Alfred A. Knopf, New York. 1977.

Gardner, W.H. and MacKenzie, N.H. *The Poems of Gerard Manley Hopkins*. Oxford University Press, Oxford. 1970.

Garrison, W.M., Morrison, D.C., Hamilton, J.G., Benson, A.A. and Calvin, M. 'Reduction of Carbon Dioxide in Aqueous Solutions by Ionizing Radiation' in *Science*. Volume 114. October 19, 1951.

Garstang, Walter. *Larval Forms and Other Zoological Verses*. Basil Blackwell, Oxford. 1951.

Garth, S. *Ovid's Metamorphoses in Fifteen Books*. The Heritage Press, New York. 1961.

Gay, John. *Rural Sports*. Printed for J. Tonson, London. 1713.

Gay, John. *The Poetical Works of John Gay*. Volume I. Charles Scribner's Sons, New York. 1893.

Gay, John. *The Poetical Works of John Gay*. Volume II. Charles Scribner's Sons, New York. 1893.

Geddes, Patrick and Thomson, J. Arthur. *Evolution*. Henry Holt and Company, New York. 1911.

Gerhard, John. *The Herball or Generall Historie of Plantes*. Printed by Adam Norton and Kichard Whitakers, London. 1633.

Gilbert, William S. *The Complete Plays of Gilbert and Sullivan*. The Modern Library, New York. 1936.

Gill, Eric. *Essays*. Jonathan Cape, London. 1947.

Gillispie, Charles Coulston. *The Edge of Objectivity*. Princeton University Press, Princeton. 1960.

Goethe, Johann Wolfgang von. *Botanical Writings Translated by Bertha Mueller*. University of Hawai'i Press, Honolulu. 1952.

Goethe, J.W. von. 'An Attempt to Interpret the Metamorphosis of Plants' in *Chronica Botanica*. Translated by Agnes Arber. Volume 10, Number 2. Summer 1946.

Goethe, Johann Wolfgang von. 'Faust' in *Great Books of the Western World*. Volume 47. Encyclopaedia Britannica, Inc., Chicago. 1952.

Goethe, Johann Wolfgang von. *Italian Journey*. Volume 6. Translated by Robert R. Heitner. Suhrkamp Publishers, New York. 1989.

Gold, Harvy J. *Mathematical Modeling of Biological Systems*. John Wiley & Sons, New York. 1977.

Goldsmith, Oliver. *The Complete Poetical Works of Oliver Goldsmith*. Henry Frowde, London. 1911.

Gombrich, E.H. *Art and Illusion*. Pantheon Books, New York. 1960.

Goodale, Elaine and Goodale, Dora Read. *All Round the Year*. G.P. Putnam's Sons, New York. 1881.

Gooday, Graeme. 'Nature in the Laboratory' in *British Journal for the History of Science*. Volume 24. 1991.

Gore, Rick. 'The Awsome Worlds Within A Cell' in *National Geographic*. Volume 150, Number 3. September 1976.

Gould, Stephen Jay. 'The Verdict on Creationism' in *The Skeptical Inquirer*. Volume 12. Winter 87/88.

Gould, Stephen Jay. 'What Color is a Zebra?' in *Natural History*. Volume 90, Number 8. August 1981.

Gould, Stephen Jay. *Ever Since Darwin*. W.W. Norton & Company, Inc., New York. 1977.

Graham, Aelred. *Christian Thought in Action*. Collins, London. 1958.

Greene, Graham. *Travels with My Aunt*. The Viking Press, New York. 1969.

Gregg, Alan. *The Furtherance of Medical Research*. Yale University Press, New Haven. 1941.

Gregory, Dick. *The Shadow That Scares Me*. Doubleday & Company, Inc., Garden City. 1968.

Gressitt, J. Linsley. *Pacific Basin Biogeography*. Bishop Museum Press, Honolulu. 1963.

Grew, Nehemiah. *The Anatomy of Plants*. Second Edition. Printed by W. Rawlins. 1682.

Grindal, Bruce and Salamone, Frank. *Bridges to Humanity*. Waveland Press, Inc., Prospect Heights. 1995.

Grobstein, Clifford. *The Strategy of Life*. W.H. Freeman and Company, San Francisco. 1974.

Haeckel, Ernst. *The Riddle of the Universe*. Harper & Brothers Publishers, New York. 1901.

Haldane, J.B.S. *The Causes of Evolution*. Longman, Green and Co., London. 1935.

Haldane, J.B.S. *What is Life?* Lindsay Drummond, London. 1949.

Haldeman-Julius, E. *Poems of Evolution*. Ten Cent Pocket Series No. 71. Haldeman-Julius Company, Girard. 1922.

Hales, Stephen. *Vegetable Staticks*. The Scientific Book Guild, London. 1961.

Halle, Louis J. *Spring in Washington*. William Sloane Associates, Inc., New York. 1947.

Handler, Philip. *Biology and the Future of Man*. Oxford University Press, New York. 1970.

Hanson, Norwood Russell. *Patterns of Discovery*. Cambridge University Press, Cambridge. 1961.

Hardy, Thomas. *Collected Poems of Thomas Hardy*. The Macmillan Company, New York. 1928.

Hardy, Thomas. *The Return of the Native*. Harper & Brothers, Publishers, New York. 1895.

Harré, Rom. *Varieties of Realism*. Basil Blackwell, Oxford. 1986.

Hartley, David. *Observations on Man*. Volume I. Printed by S. Richardson, London. 1749.

Haught, James A. *2000 Years of Disbelief*. Prometheus Books, Amherst. 1996.

Hawthorne, Nathaniel. *Mosses from an Old Manse*. Henry Altemus Compert, Philadelphia. 1900.

Hay, John. *The Undiscovered Country*. W.W. Norton & Company, New York. 1981.

Hayne, Paul Hamilton. *Sonnet*. 1860.

Hays, H.R. *Birds, Beasts, and Men*. G.P. Putnam's Sons, New York. 1972.

Hecht, Max K. *Evolutionary Biology at the Crossroads*. Queens College Press, Flushing. 1989.

Hediger, Heini. *Man and Animal in the Zoo.* Translated by Gwynne Vevers and Winwood Reade. Delacorte Press, New York. 1969.

Heine, Heinrich. *The Book of Songs.* The Roycrofters, East Aurora. 1903.

Heinlein, Robert A. *Time Enough for Love.* G.P. Putnam's Sons, New York. 1973.

Hemans, Felicia. *The Poetical Works of Mrs. Felicia Hemans.* William Collins, Sons, & Co., Limited, London. nd.

Hemingway, Ernest. *The Old Man and the Sea.* Jonathan Cape, London. 1952.

Henderson, Lawrence J. *The Fitness of the Environment.* The Macmillan Company, New York. 1913.

Henley, William. *Echoes of Life and Death.* Thomas B. Mosher, Portland. 1908.

Henslow, John Stevens. 'On the Requisites Necessary for the Advance of Botany' in *Magazine of Zoology and Botany.* Volume 1. 1837.

Herford, Oliver. *A Child's Primer of Natural History.* Charles Scribner's Sons, New York. 1899.

Herzberg, Max J. *Narrative Poems.* D.C. Heath and Company, Boston. 1930.

Hillyard, Paul. *The Book of the Spider.* Random House, New York. 1994.

Himmelfarb, Gertrude. *Darwin and the Darwinian Revolution.* Chatto & Windus, London. 1959.

Hine, Reginald L. *Confessions of an Un-Common Attorney.* The Macmillan Company, New York. 1949.

Hitching, Francis. *The Neck of the Giraffe.* Ticknor & Fields, New Haven. 1982.

Hoage, R.J. *Animal Extinctions.* Smithsonian Institution Press, Washington, D.C. 1985.

Hobbes, Thomas. 'Leviathan' in *Great Books of the Western World.* Volume 23. Encyclopaedia Britannica, Inc., Chicago. 1952.

Hofstadter, Richard. *Anti-Intellectualism in American Life.* Alfred A. Knopf, New York. 1963.

Holland, W.J. *The Moth Book.* Doubleday, Page & Company, New York. 1904.

Hölldobler, Bert and Wilson, Edward O. *Journey to the Ants: A Story of Scientific Exploration.* Harvard University Press, Cambridge. 1994.

Holmes, Bob. 'Life is...?' in *New Scientist.* Number 2138. 13 June 1998.

Holmes, Oliver Wendell. 'The Professor at the Breakfast Table' in *More Yamkee Drolleries.* John Camden Hotten, Piccadilly, London. 1869.

Holmes, Oliver Wendell. *The Complete Poetical Works of Oliver Wendell Holmes.* Houghton, Mifflin and Company, Boston. 1899.

Holmes, Oliver W. *The Poet at the Breakfast-Table.* Houghton, Mifflin and Company, Boston. 1892.

Hood, Thomas. *The Poetical Works of Thomas Hood.* William Collins, Sons & Co., Ltd, London. nd.

Hooke, Robert. *Micrographia*. Printed for the Royal Society. 1665.

Hopwood, A.T. 'The Development of Pre-Linnaean Taxonomy' in *Proceedings of the Linnean Society of London*. Volume 170. 1959.

Horace. *Satires, Epistles and Ars Poetica*. Translated by H. Rushton Fairclough. Harvard University Press, Cambridge. 1964.

Hornaday, William T. *Our Vanishing Wild Life*. Arno Press, New York. 1970.

Horsfield, Brenda and Stone, Peter Bennet. *The Great Ocean Business*. Coward, McCann & Geoghegan Inc., New York. 1972.

Hughes, T.E. *Mites or the Acari*. The Athlone Press, University of London. 1959.

Hugo, Victor. *Les Miserables*. New American Library, New York. 1987.

Hull, D.L. 'Individuality and Selection' in *Annual Review of Ecology and Systematics*. Volume 11. 1980.

Hume, David. *An Enquiry Concerning Human Understanding*. Oxford University Press, Oxford. 1999.

Hunt, Leigh. *The Poetical Works of Leigh Hunt*. Routledge, Warner, Routledge, London. 1860.

Hurdis, James. *Sir Thomas More: A Tragedy*. Printed for J. Johnson, London. 1793.

Hutchins, Robert M. and Mortimer, J. Adler. *The Great Ideas of Today 1974*. Encyclopaedia Britannica Inc., Chicago. 1974.

Hutton, J. 'Theory of the Earth' in *Transactions of the Royal Society of Edinburgh*. Volume 1. 1788.

Huxley, Aldous. *Brave New World*. Harper & Brothers Publishers, New York. 1946.

Huxley, Aldous. *Time Must Have A Stop*. Harper & Brothers Publishers, New York. 1944.

Huxley, Julian. *Essays of a Biologist*. Chatto & Windus, London. 1923.

Huxley, Julian. *Essays in Popular Science*. Alfred A. Knopf, New York. 1927.

Huxley, Julian. *Evolution: The Modern Synthesis*. George Allen & Unwin Ltd, London. 1942.

Huxley, Julian. *Evolution in Action*. Harper & Row, New York. 1953.

Huxley, Julian. *Memories*. George Allen and Unwin Ltd, London. 1970.

Huxley, Julian. *The Captive Shrew*. Harper & Brothers Publishers, New York. 1933.

Huxley, Julian. *The New Systematics*. Clarendon Press, Oxford. 1940.

Huxley, Leonard. *Life and Letters of Thomas Henry Huxley*. Volume 1. Macmillan and Co., Ltd, London. 1903.

Huxley, Thomas. *Collected Essays*. Volume II. Macmillan and Company, London. 1893.

Huxley, Thomas. *Collected Essays*. Volume III. Macmillan and Company, London. 1893.

Huxley, Thomas. *Collected Essays*. Volume VI. Macmillan and Company, London. 1893.

Huxley, Thomas. *Collected Essays*. Volume VII. Macmillan and Company, London. 1893.

Huxley, Thomas. *Evolution & Ethics*. Macmillan and Co., London. 1894.

Huxley, Thomas. *Lay Sermons, Addresses, and Reviews*. D. Appleton & Company, New York. 1871.

Ingelow, Jean. *Poems*. Roberts Brothers, Boston. 1863.

Ingelow, Jean. *Songs of Seven*. Roberts Brothers, Boston. 1883.

Irvine, William. *Apes, Angels, and Victorians*. McGraw-Hill Book Company, Inc., New York. 1955.

Jackson, Helen. *Verses*. Roberts Brothers, Boston. 1890.

Jacob, François. *The Possible and the Actual*. Pantheon Books, New York. 1982.

Jaffe, Bernard. *Crucibles*. Tudor Publishing Co., New York. 1937.

James, Henry. *The Letters of William James*. Volume II. The Atlantic Monthly Press, Boston. 1920.

James, William. *The Principles of Psychology*. Volume II. Henry Holt and Company, New York. 1890.

James, William. *The Varieties of Religious Experience*. Harvard University Press, Cambridge. 1985.

James, William. *The Will to Believe and Other Popular Essays in Popular Philosophy*. Harvard University Press, Cambridge. 1979.

Jaroff, L. 'The Gene Hunt' in *Time*. Volume 133, Number 11. March 20, 1989.

Jeffreys, Harold. *Theory of Probability*. Third edition. Clarendon Press, Oxford. 1967.

Jennings, H.S. *The Universe and Life*. Yale University Press, New Haven. 1941.

Johanson, Donald C and Edey, Maitland A. *Lucy: The Beginnings of Humankind*. Warner Books, New York. 1981.

Johnson-Laird, P.N. *The Computer and the Mind*. Harvard University Press, Cambridge. 1988.

Johnson, Samuel. *A Journey to the Western Islands of Scotland*. Printed by Thomas Walker, Dublin. 1775.

Johnson, Samuel. *Rambler*. Volume I. Garland Publishing, Inc., New York. 1978.

Jones, H. Spencer. *Life on Other Worlds*. The English Universities Press, Ltd, London. 1940.

Joseph, Lawrence E. *GAIA: The Growth of an Idea*. St Martin's Press, New York. 1990.

Judson, Horace Freeland. *The Eighth Day of Creation*. Simon and Schuster, New York. 1979.

Jukes, Thomas H. *Molecules and Evolution*. Columbia University Press, New York. 1966.

Kaku, Michio and Trainer, Jennifer. *Beyond Einstein*. Bantam Books, Toronto. 1987.

Kames, Henry. *The Gentleman Farmer*. Printed for W. Creech, Edinburgh. 1776.

Kant, Immanuel. *Critique of Pure Reason*. Translated by Norman Kemp Smith. St Martin's Press, New York. 1965.

Kant, Immanuel. *Universal Natural History and Theory of the Heavens*. The University of Michigan Press, Ann Arbor. 1969.

Kauffman, Stuart. *At Home in the Universe*. Oxford University Press, New York. 1995.

Keats, John. *Complete Poems*. Harvard University Press, Cambridge. 1982.

Kellog, Vernon. 'The Biologist Speaks of Death' in *The Atlantic Monthly*. June 1921.

Kendall, May. *Dreams to Sell*. Longman, Green & Co., London. 1887.

Key, Archie F. *Beyond Four Walls: The Origins and Development of Canadian Museums*. McClelland and Stewart Limited, Toronto. 1973.

Keynes, Geoffrey and Hill, Brian. *Samuel Butler's Notebooks*. E.P. Dutton & Company, Inc., Publishers, New York. 1950.

Kirby, William and Spence, William. *An Introduction to Entomology*. Longman, Green, Longman, and Roberts, London. 1860.

Kitcher, Philip. *Abusing Science*. The MIT Press, Cambridge. 1982.

Klee, Paul. *The Inward Vision*. Harry N. Abrahams, Inc., New York. 1958.

Kluckhohn, Clyde. *Mirror for Man*. McGraw-Hill Book Company, Inc., New York. 1949.

Knight, Damon. *Charles Fort: Prophet of the Unexplained*. Victor Gollancz Ltd, London. 1971.

Koestler, Arthur. *Janus: A Summing Up*. Hutchinson of London, London. 1978.

Koestler, Arthur. *The Act of Creation*. The Macmillan Company, New York. 1964.

Koestler, Arthur and Smythies, J.R. *Beyond Reductionism*. Beacon Press, Boston. 1968.

Kornberg, Arthur. 'The Two Cultures: Chemistry and Biology' in *Biochemistry*. Volume 26, Number 22. November 3, 1987.

Krause, Ernst. *Erasmus Darwin*. John Murray, London. 1879.

Krebs, H.A. 'How the Whole Becomes More than the Sum of the Parts' in *Perspectives in Biology and Medicine*. Volume 14, Number 3. Spring 1971.

Krutch, Joseph Wood. *The Desert Year*. William Sloane Associates, New York. 1952.

Krutch, Joseph Wood. *The Great Chain of Life*. Houghton Mifflin Company, Boston. 1957.

Krutch, Joseph Wood. *The Modern Temper*. Harcourt, Brace and Company, New York. 1929.

Krutch, Joseph Wood. *The Twelve Seasons*. William Sloane Publishers, New York. 1949.

Kuhn, Thomas S. *The Structure of Scientific Revolutions*. The University of Chicago Press, Chicago. 1970.

Kurlansky, Mark. *Cod*. Walker and Company, New York. 1997.

Lacinio, Giano. *The New Pearl of Great Price*. Arno Press, New York. 1974.

Laglands, R. *Biographical Memoirs of Fellows of the Royal Society*. The Royal Society, London. 1985.

Laird, John. *A Study in Realism*. Cambridge University Press, Cambridge. 1920.

Lamarck, J.B. *Hydrogeology*. University of Illinois Press, Urbana. 1964.

Lamarck, J.B. *Zoological Philosophy*. The University of Chicago Press, Chicago. 1984.

Landau, H.G. 'Mathematical Biology' in *Science*. Volume 114.

Landon, L.E. *The Poetical Works of Miss Landon*. Henry F. Anners, Philadelphia. 1842.

Langer, Susanne K. *Philosophy in a New Key*. Harvard University Press, Cambridge. 1961.

Lanier, Sidney. *The Marshes of Glynn*. The Ashantilly Press, Darien. 1967.

Large, E.C. *The Advance of the Fungi*. Henry Holt & Company, New York. 1940.

Latour, Bruno. *Science in Action*. Open University Press, Milton Keynes. 1987.

Laudan, Larry. 'Commentary: Science at the Bar—Cause for Concern' in *Science, Technology & Human Values*. Volume 7, Number 41. Fall 1982.

Lavoisier, Antoine. 'Elements of Chemistry' in *Great Books of the Western World*. Volume 45. Encyclopaedia Britannica, Inc., Chicago. 1952.

Lawrence, D.H. *Fantasia of the Unconscious*. Thomas Seltzer, New York. 1922.

Lawrence, D.H. *The Complete Poems of D.H. Lawrence*. Volume I. The Viking Press, New York. 1964.

Lawrence, D.H. *The Complete Poems of D.H. Lawrence*. Volume II. The Viking Press, New York. 1964.

Lawrence, Jerome, and Lee, Robert E. *The Night Thoreau Spent in Jail*. Hill & Wang, New York. 1970.

Lear, Edward. *Of Pelicans and Pussycats*. Dial Books for Young Readers, New York. 1990.

Leffler, John E. and Grunwald, Ernest. *Rates and Equilibria of Organic Reactions*. John Wiley and Sons, Inc., New York. 1963.

Leland, Charles Godfrey. *The Music-Lesson of Confucius.* James R. Osgood and Company, Boston. 1872.

Leob, Jacques. 'The Recent Development of Biology' in *Science.* Volume 20. 1904.

Leob, J. *The Mechanistic Conception of Life.* The University of Chicago Press, Chicago. 1912.

Leopold, Aldo. *A Sand County Almanac: With Essays on Conservation from Round River.* Oxford University Press, Oxford. 1966

Lewis, C.S. *Studies in Words.* Cambridge University Press, Cambridge. 1960.

Lewis, Wyndham. *The Caliph's Design.* Black Sparrow Press, Santa Barbara. 1986.

Liebig, Justus. *Animal Chemistry.* Johnson Reprint Corporation, New York. 1964.

Lindburg, Charles A. 'The Wisdom of Wilderness' in *Life.* Volume 63, Number 25. December 22, 1967.

Lindsay, Vachel. *Collected Poems.* The Macmillan Company, New York. 1923.

Linne, Carl von. *Critica Botanica.* Translated by Sir Arthur Hort. Printed for the Royal Society, London. 1938.

Loewy, Ariel G. and Siekevitz, Philip. *Cell Structure and Function* Second Edition. Holt, Reinhart and Winston, Inc., New York. 1969.

London, Jack. *The Sea-Wolf.* Houghton Mifflin Company, Boston. 1964.

Longfellow, Henry Wadsworth. *The Complete Writings of Henry Wadsworth Longfellow.* Volume I. Houghton Mifflin Company, Boston. 1904.

Longfellow, Henry Wadsworth. *The Complete Writings of Henry Wadsworth Longfellow.* Volume II. Houghton Mifflin Company, Boston. 1904.

Longfellow, Henry Wadsworth. *The Complete Writings of Henry Wadsworth Longfellow.* Volume III. Houghton Mifflin Company, Boston. 1904.

Longfellow, Henry Wadsworth. *The Complete Writings of Henry Wadsworth Longfellow.* Volume IV. Houghton Mifflin Company, Boston. 1904.

Longfellow, Henry Wadsworth. *The Complete Writings of Henry Wadsworth Longfellow.* Volume V. Houghton Mifflin Company, Boston. 1904.

Longfellow, Henry Wadsworth. *The Complete Writings of Henry Wadsworth Longfellow.* Volume VI. Houghton Mifflin Company, Boston. 1904.

Lote, Christopher J. 'Correspondence' in *Nature.* Volume 363, Number 6428. 3 June 1993.

Lovejoy, Thomas E. 'The Quiet Apocalypse' in *Time.* Volume 128, Number 15. October 1968.

Lovelace, Richard. *The Poems of Richard Lovelace.* Clarendon Press, Oxford. 1968.

Lovelock, J.E. *Gaia.* Oxford University Press, Oxford. 1979.

Lover, Samuel. *Poems of Ireland.* Ward, Lock & Baldwin, Limited, London. 1893.

Lowell, Amy. *The Complete Poetical Works of Amy Lowell.* The Riverside Press, Cambridge. 1955.

Lowell, Maria. *The Poems of Maria Lowell.* The Riverside Press, Cambridge. 1907.

Lowman, Margaret D. *Life in the Treetops.* Yale University Press, New Haven. 1999.

Lubbock, John. *The Beauties of Nature.* Macmillan and Co., Limited, London. 1900.

Lueders, Edward. *Writing Natural History.* University of Utah Press, Salt Lake City. 1989.

Luria, S.E. *A Slot Machine, A Broken Test Tube.* Harper & Row, Publishers. New York. 1984.

Lynd, Robert. *The Blue Lion.* Books for Libraries Press, Freeport. 1968.

Lynd, Robert. *The Peal of Bells.* Methuen & Co., Ltd, London. 1927.

Lyon, Jeff, and Corner, Peter. *Altered Fates.* W.W. Norton, New York. 1995.

MacArthur, Robert H. *Geographical Ecology.* Harper & Row, Publishers, New York. 1972.

MacDonald, George. *The Poetical Works of George MacDonald.* Chatto & Windus, London. 1915.

Mach, Ernst. *The Science of Mechanics.* The Open Court Publishing Co., La Salle. 1942.

MacLeish, Archibald. *The Collected Poems of Archibald MacLeish.* Houghton Mifflin Company, Boston. 1962.

Madách, Imre. *The Tragedy of Man.* Dr. George Vajna & Co., Budapest. 1933.

Maddox, John. 'The Dark Side of Molecular Biology' in *Nature.* Volume 363. 6 May 1993.

Mallis, Arnold. *American Entomologists.* Rutgers University Press, New Brunswick. 1971.

Mann, Thomas. *The Magic Mountain.* Alfred A. Knopf, New York. 1958.

Marquis, Don. *The Lives and Times of Archy & Mehitabel.* Doubleday Doran & Co., Inc., Garden City. 1933.

Marsh, George Perkins. *Man and Nature; or, Physical Geography as Modified by Human Action.* Harvard University Press, Cambridge. 1965.

Martin, Charles Noël. *The Role of Perception in Science.* Translated by A.J. Pomerans. Hutchinson of London, London. 1963.

Mather, K.F. 'Forty Years of Scientific Thought Concerning the Origin of Life' in *Journal of the Scientific Laboratories of Denison University.* Volume 22. 1927.

Mayr, Ernst. *Systematics and the Origin of Species.* Columbia University Press, New York. 1942.

Mayr, Ernst. *Populations, Species, and Evolution.* Harvard University Press, Cambridge. 1970.

Mayr, Ernst. *The Growth of Biological Thought: Diversity, Evolution, Inheritance*. Harvard University Press, Cambridge. 1982.

Mayr, Ernst and Provine, William B. *The Evolutionary Synthesis*. Harvard University Press, Cambridge. 1980.

McArthur, Peter. *The Best of Peter McArthur*. Clarke, Irwin & Company, Limited, Toronto. 1967.

McKibben, Bill. *The End of Nature*. Random House, New York. 1989.

Mead, George Herbert. *The Philosophy of the Act*. The University of Chicago Press, Chicago. 1938.

Medawar, Peter. *Pluto's Republic*. Oxford University Press, Oxford. 1982.

Medawar, Peter B. *The Hope of Progress*. Anchor Press, Garden City. 1973.

Medawar, P.B. and Medawar, J.S. *The Life Science*. Harper & Row, Publishers, New York. 1977.

Menzel, Donald H. and Boyd, Lyle G. *The World of Flying Saucers*. Doubleday & Company, Inc., Garden City. 1963.

Meyerson, Emile. *Identity & Reality*. George Allen & Unwin Ltd, London. 1930.

Midgley, M. 'Gene-juggling' in *Philosophy*. Volume 54, Number 210. 1979.

Midgley, Mary. 'Can Science Save Its Soul?' in *New Scientist*. Volume 135, Number 1832. 1 August 1992.

Mill, John Stuart. *Three Essays on Religion*. AMS Press, New York. 1970.

Millay, Edna St Vincent. *Collected Poems*. Harper & Row, Publishers, New York. 1917.

Millay, Edna St Vincent. *Collected Sonnets*. Harper & Row, Publishers, New York. 1988.

Miller, Douglas. *Scientific Studies*. Volume 12. Suhrkamp Publishers New York, New York. 1988.

Miller, Sally M. *John Muir: Life and Work*. University of New Mexico Press, Albuquerque. 1993.

Milne, A.A. *Toad of Toad Hall*. Samuel French Ltd, New York. 1947.

Milton, John. 'Lycidas' in *Great Books of the Western World*. Volume 32. Encyclopaedia Britannica, Inc., Chicago. 1952.

Milton, John. 'Paradise Lost' in *Great Books of the Western World*. Volume 32. Encyclopaedia Britannica, Inc., Chicago. 1952.

Minnaert, M. *The Nature of Light & Colour in the Open Air*. Dover Publications Inc. 1954.

Mishima, Yukio. *The Sound of the Waves*. Berkley Publishing Corporation, New York. 1961.

Moffett, Thomas. *The Theater of Insects*. Printed by E.C., London. 1658.

Moir, D.M. *The Poetical Works of David Macbeth Moir*. William Blackwood and Sons, Edinburgh. 1852.

Monod, Jacques. *Chance and Necessity*. Vintage Books, New York. 1972.

Montagu, Ashley. *Science and Creationism*. Oxford University Press, Oxford. 1984.

Montagu, George. *Testacea Britannica*. Printed by J.S. Hollis, London. nd.

Montaigne, Michel Eyquem de. 'The Essays' in *Great Books of the Western World*. Volume 25. Encyclopaedia Britannica, Inc., Chicago. 1952.

Montgomery, James. *Poetical Works of James Montgomery*. Volume II. Philips, Sampson and Company, Boston. 1859.

Moore, Ruth. *The Coil of Life*. Alfred A. Knopf, New York. 1961.

Moore, Thomas. *The Poetical Works of Thomas Moore*. The World Publishing House, New York. 1877.

Mora, P.T. 'Urge and Molecular Biology' in *Nature*. Volume 199, Number 4890. July 20, 1963.

Morgan, Thomas Hunt. *The Physical Basis of Heredity*. J.B. Lippincott Company, Philadelphia. 1919.

Morley, John Viscount. *Critical Miscellanies*. Macmillan and Co., Limited, London. 1923.

Morris, Desmond. *The Naked Ape*. McGraw-Hill Book Company, New York. 1967.

Morris, George P. *Poems*. Charles Scribner, New York. 1854.

Morris, Henry M. *Scientific Creationism*. Creation-Life Publishers, San Diego. 1974.

Morton, Ron L. *Music of the Earth*. Plenum Press, New York. 1996.

Moss, Ralph W. *Free Radical*. Paragon House Publishers, New York. 1988.

Muir, John. *Gentle Wilderness*. Ballantine Books. 1968.

Muir, John. *A Thousand-Mile Walk to the Gulf*. Houghton Mifflin Company, Boston. 1916.

Muller, H.J. 'Life' in *Science*. Volume 121. 7 January, 1955.

Muller, H.J. 'How Radiation Changes the Genetic Constitution' in *Bulletin of the Atomic Scientists*. Volume XI, Number 9. November 1955.

Muller, Herbert J. *Science and Criticism*. Yale University Press, New Haven. 1943.

Mulock, Dinah Maria. *Miss Mulock's Poems*. Houghton, Osgood, and Company, Boston. 1880.

Murchie, Guy. *The Seven Mysteries of Life*. Houghton Mifflin Company, Boston. 1978.

Murry, J. Middleton. *Journal of Katherine Mansfield*. Alfred A. Knopf, New York. 1946.

Museums Association. 'The Principles of Museum Administration' in *Report of Proceedings*. 1895.

Myers, Norman. *A Wealth of Wild Species*. Westview Press, Boulder. 1983.

Nagle, James J. 'Genetic Engineering' in *Bulletin of the Atomic Scientists*. December 1971.

Nash, Ogden. *Verses from 1929 On*. Little, Brown and Company, Boston. 1959.

Nash, Ogden. *Everyone But Thee and Me*. Little, Brown and Company, Boston. 1962.

Needham, James G. 'Developments in Philosophy of Biology' in *Quarterly Review of Biology*. Volume III, Number 1. March 1928.

Needham, Joseph. *Order and Life*. Yale University Press, New Haven. 1936.

Needham, Joseph. *Time: The Refreshing River*. The Macmillan Company, New York. 1943.

Nelkin, Dorothy. *Science Textbook Controversies and the Politics of Equal Time*. The MIT Press, Cambridge. 1977.

Newman, J.R. *What is Science*. Simon and Schuster, New York. 1955.

Newman, Joseph H. *Poems for Penguins*. Greenberg Publishers, New York. 1941.

Newton, Sir Isaac. 'Optics' in *Great Books of the Western World*. Volume 34. Encyclopaedia Britannica, Inc., Chicago. 1952.

Nicholson, Norman. *A Local Habitation*. Faber and Faber, London. 1972.

Oemler, Marie Conway. *Slippy Magee, Sometimes Known as Butterfly Man*. Grosset & Dunlap Publishers, New York. 1917.

Oliver, Bernard M. *Project Cyclops*. Moffett Field, California. 1973.

Oliver, Mary. *Blue Pastures*. Harcourt Brace & Company, New York. 1991.

Olson, S.L. 'The Museum Tradition in Ornithology. A Response to Ricklefs' in *The Auk*. Volume 98. January 1981.

Orgel, Irene. *The Odd Tales of Irene Orgel*. The Eakins Press, New York. 1966.

Osler, Sir William. *Aequanimitas*. The Blakistan Company, Philadelphia. 1942.

Outwater, Alice. *Water: a Natural History*. Basic Books, New York. 1996.

Page, Jake. *Pastorale*. W.W. Norton & Company, New York. 1985.

Pallister, William. *Poems of Science*. Playford Press, New York. 1931.

Pascal, Blaise. 'Pensées' in *Great Books of the Western World*. Volume 33. Encyclopaedia Britannica, Inc., Chicago. 1952.

Pauling, Linus. *No More War!*. Greenwood Press, Publishers, Westport, CT. 1962.

Peacock, A.R. 'Reductionism: a Review of the Epistemological Issues and Their Relevance in Biology and the Problem of Consciousness' in *Zygon*. Volume 11, Number 4. 4 December 1976.

Peacock, Thomas Love. *Gryll Grange*. Penguin Books, Harmondsworth. 1949.

Pearson, Karl. 'Mathematics and Biology' in *Nature*. Volume 63, Number 1629. January 17, 1901.

Pearson, Karl. *The Grammar of Science*. Part I. 1911.

Peattie, Donald Culross. *An Almanac for Moderns*. G.P. Putnam's Sons, New York. 1935.

Peattie, Donald Culross. *Flowering Earth*. G.P. Putnam's Sons, New York. 1939.

Peattie, Donald Culross. *The Road of a Naturalist*. Robert Hale Limited, London. 1946.

Peirce, Charles, Sanders. *The Collected Papers of Charles Sanders Peirce*. Volume VI. Harvard University Press, Cambridge. 1935.

Pepper, Stephen C. *World Hypotheses*. University of California Press, Berkeley. 1970.

Perrin, Noel. *Second Person Rural*. David R. Godine, Publisher, Boston. 1980.

Peters, Ted. *Playing God?* Routledge, New York. 1997.

Plath, Sylvia. *The Collected Poems*. Harper Perennial, New York. 1992.

Playfair, John. *Illustrations of the Huttonian Theory of the Earth*. Printed for Cadell and Davies. Edinburgh. 1802.

Poe, Edgar Allan. *Little Masterpieces*. Double Day & McClure Co., New York. 1897.

Poe, Edgar Allan. *The Raven and Other Poems*. Columbia University Press, New York. 1942.

Poincare, Henri. *The Foundations of Science*. The Science Press, Lancaster. 1946.

Polanyi, Michael. *Personal Knowledge*. Harper & Row, Publishers, New York. 1962.

Pope, Alexander. *Alexander Pope's Collected Poems*. Everyman's Library, New York. 1965.

Popper, Karl R. *Objective Knowledge: An Evolutionary Approach*. Clarendon Press, Oxford. 1983.

Popper, Karl R. *The Logic of Scientific Discovery*. Harper & Row, Publishers, New York. 1968.

Pownall, Thomas. *A Topographical Description of the Dominions of the United States of America*. University of Pittsburgh Press, Pittsburgh. 1949.

Pratchett, Terry. *Equal Rites*. Victor Gollancz Ltd, London. 1987.

Preble, Edward A. *Nature Magazine*. Volume 50, Number 10. December 1957.

Prelutsky, Jack. *A Pizza the Size of the Sun*. Greenwillow Books, New York. 1996.

Prelutsky, Jack. *Something Big Has Been Here*. Greenwillow Books, New York. 1990.

Price, Lucien. *Dialogues of Alfred North Whitehead*. Little, Brown and Company, Boston. 1954.

Pringle, Thomas. *Afar in the Desert*. Longmans, Green, and Co., London. 1881.

Proust, Marcel. *Pleasures and Regrets*. Translated by Louise Varese. Crown Publishers, New York. 1948.

Provine, William B. *Sewall Wright and Evolutionary Biology*. The University of Chicago Press, Chicago. 1986.

Purcell, Rosamond and Gould, Stephen Jay. *Illiminations: A Bestiary*. W.W. Norton & Company, New York. 1986.

Purchas the Younger, Samuel. *A Theatre of Political Flying-Insects*. Printed by R.I. for Thomas Parkhurst, London. 1657.

Quammen, David. *Natural Acts*. Nick Lyons Books, New York. 1985.

Queneau, Raymond. *Exercises in Style*. New Direction Publishing Corporation, New York. 1947.

Ramsey, Paul. *Faith and Ethics*. Harper & Brothers, New York. 1957.

Raymo, Chet. *The Virgin and the Mousetrap*. Viking, New York. 1991.

Reade, Winwood. *The Martyrdom of Man*. Kegan Paul, Trench, Trubner & Co., Ltd, London. 1910.

Reichenbach, Hans. *The Rise of Scientific Philosophy*. University of California Press, Berkeley. 1951.

Richet, Charles. *The Natural History of a Savant*. Translated by Sir Oliver Lodge. J.M. Dent & Sons Limited, London. 1927.

Richter, Jean Paul. *The Literary Works of Leonardo Da Vinci*. Volume II. Oxford University Press, London. 1939.

Ricketts, Edward F., Calvin, Jack and Hedgpeth, Joel W. *Between Pacific Tides*. Stanford University Press, Stanford. 1985.

Riley, James Whitcomb. *The Complete Works of James Whitcomb Riley*. Volume III. P.F. Collier & Son, Company, New York. 1916.

Riley, James Whitcomb. *The Complete Works of James Whitcomb Riley*. Volume IV. P.F. Collier & Son, Company, New York. 1916.

Riley, James Whitcomb. *The Complete Works of James Whitcomb Riley*. Volume VII. P.F. Collier & Son, Company, New York. 1916.

Riley, James Whitcomb. *The Complete Works of James Whitcomb Riley*. Volume VIII. P.F. Collier & Son, Company, New York. 1916.

Roberts, Catherine. 'The Use of Animals in Medical Research—Some Ethical Considerations' in *Perspectives in Biology and Medicine*. Volume VIII, Number 1. Autumn 1964.

Roberts, Catherine. *Science, Animals, and Evolution*. Greenwood Press, Westport. 1980.

Robson, Ann P. and Robson, John M. *Collected Works of John Stuart Mill*. Volume XXII. University of Toronto Press, Toronto. 1963.

Roe, Anne and Simpson, George Gaylord. *Behavior and Evolution*. Yale University Press, New Haven. 1958.

Rohlfing, Duane L. and Oparin, A.I. *Molecular Evolution: Prebiological and Biological*. Plenum Press, New York. 1972.

Rollins, R.C. 'Taxonomy of the Higher Plants' in *American Journal of Botany*. Volume 44, Number 1. January 1957.

Root, R.K. 'The Age of Faith' in *The Atlantic Monthly*. Volume CX. July 1912.

Rossetti, Christina G. *The Complete Poems of Christina Rossetti*. Volume I. Louisiana State University Press, Baton Rouge. 1979.

Rostand, Jean. *Can Man Be Modified?* Basic Books, Inc., New York. 1959.

Rostand, Jean. *Humanly Possible*. Translated by Lowell Bair. Saturday Review Press, New York. 1973.

Rubin, Harry. 'Cancer as a Dynamic Developmental Disorder' in *Cancer Research*. Volume 45. July 1985.

Rudloe, Jack. *Time of the Turtle*. Alfred A. Knopf, New York. 1979.

Ruskin, John. *The Queen of the Air*. The Macmillan Company, New York. 1928.

Russell, Bertrand. *Mysticism and Logic*. Doubleday & Company, Inc., Garden City. 1950.

Rutter, R.J. *W.E. Saunders—Naturalist*. The University of Toronto Press, Toronto. 1949.

Sackville-West, V. *All Passion Spent*. Doubleday, Doran & Company, Inc., Garden City. 1931.

Sakaki, Nonao. *Break the Mirror*. North Point Press, San Francisco. 1987.

Salam, Abdus. 'The Role of Chirality in the Origin of Life' in *Journal of Molecular Evolution*. Volume 33, Number 2. August 1992.

Samuel, Herbert. *Book of Quotations*. James Barrie, London. 1954.

Sanborne, Kate. 'Studies of Animal Nature' in *Atlantic Monthly*. February 1877.

Santayana, George. *Dominations and Powers*. Charles Scribner's Sons, New York. 1951.

Sappho. *Poems and Fragments*. Carol Publishing Group, New Jersey. 1993.

Savage, Jay M. *Evolution*. Holt, Rinehart and Winston, Inc., New York. 1969.

Savory, Theodore. *Naming the Living World*. The English University Press Ltd, London. 1962.

Schaefer, Jack. 'Interview with a Shrew' in *Audubon*. Volume 77, Number 6. November 1975.

Schaller, George B., Hu Jinchu, Pan Wenshi and Zhu Jing. *The Giant Pandas of Wolong*. The University of Chicago Press, Chicago. 1985.

Schenk, Edward T. and McMasters, John H. *Procedure in Taxonomy*. Stanford University Press, Stanford. 1948.

Schmidt-Nielsen, Knut. 'Scaling in Biology: The Consequence of Size' in *Journal of Experimental Zoology*. Volume 194. October, November, December 1975.

Schrödinger, Erwin. *Science and Humanism*. Cambridge University Press, Cambridge. 1952.

Schrödinger, Erwin. *What is Life?* Cambridge University Press, Cambridge. 1946.

Schweitzer, Albert. *Civilization and Ethics*. Adam & Charles Black, London. 1949.

Scientific American. *The Physics and Chemistry of Life*. Simon and Schuster, New York. 1955.

Scott, Walter. *The Lord of the Isles*. James Ballantyne and Co., Edinburgh. 1815.

Sears, Paul B. *Deserts on the March*. University of Oklahoma Press, Norman. 1935.

Seward, A.C. *Darwin and Modern Science*. Cambridge University Press, Cambridge. 1909.

Seward, A.C. *Links with the Past in the Plant World*. Cambridge University Press, Cambridge. 1921.

Shakespeare, William. 'A Midsummer-Night's Dream' in *Great Books of the Western World*. Volume 26. Encyclopaedia Britannica, Inc., Chicago. 1952.

Shakespeare, William. 'Anthony and Cleopatra' in *Great Books of the Western World*. Volume 27. Encyclopaedia Britannica, Inc., Chicago. 1952.

Shakespeare, William. 'Cymbeline' in *Great Books of the Western World*. Volume 27. Encyclopaedia Britannica, Inc., Chicago. 1952.

Shakespeare, William. 'Hamlet, Prince of Denmark' in *Great Books of the Western World*. Volume 27. Encyclopaedia Britannica, Inc., Chicago. 1952.

Shakespeare, William. 'King Lear' in *Great Books of the Western World*. Volume 27. Encyclopaedia Britannica, Inc., Chicago. 1952.

Shakespeare, William. 'Macbeth' in *Great Books of the Western World*. Volume 27. Encyclopaedia Britannica, Inc., Chicago. 1952.

Shakespeare, William. 'Measure for Measure' in *Great Books of the Western World*. Volume 27. Encyclopaedia Britannica, Inc., Chicago. 1952.

Shakespeare, William. 'Pericles, Prince of Tyre' in *Great Books of the Western World*. Volume 27. Encyclopaedia Britannica, Inc., Chicago. 1952.

Shakespeare, William. 'The First Part of King Henry the Fourth' in *Great Books of the Western World*. Volume 26. Encyclopaedia Britannica, Inc., Chicago. 1952.

Shakespeare, William. 'The Life of King Henry the Fifth' in *Great Books of the Western World*. Volume 26. Encyclopaedia Britannica, Inc., Chicago. 1952.

Shakespeare, William. 'The Merchant of Venice' in *Great Books of the Western World*. Volume 26. Encyclopaedia Britannica, Inc., Chicago. 1952.

Shakespeare, William. 'The Second Part of King Henry the Fourth' in *Great Books of the Western World*. Volume 26. Encyclopaedia Britannica, Inc., Chicago. 1952.

Shakespeare, William. 'The Taming of the Shrew' in *Great Books of the Western World*. Volume 26. Encyclopaedia Britannica, Inc., Chicago. 1952.

Shakespeare, William. 'The Tragedy of King Richard the Third' in *Great Books of the Western World*. Volume 26. Encyclopaedia Britannica, Inc., Chicago. 1952.

Shakespeare, William. 'The Winter's Tale' in *Great Books of the Western World*. Volume 27. Encyclopaedia Britannica, Inc., Chicago. 1952.

Shakespeare, William. 'Troilus and Cressida' in *Great Books of the Western World*. Volume 27. Encyclopaedia Britannica, Inc., Chicago. 1952.

Shapere, Dudley. *Philosophical Problems of Natural Science*. The Macmillan Company, New York. 1965.

Shapiro, Harry L. 'Symposium on the History of Anthropology' in *American Anthropologist*. Volume 61, Number 3. 1959.

Shaw, George Bernard. *Back to Methuselah*. Constable and Company, Ltd, London. 1921.

Shaw, George Bernard. *Man and Superman*. The University Press, Cambridge. 1903.

Sheldrick, Daphne. *The Tsavo Story*. Collins and Harvill Press, London. 1973.

Shelley, Percy Bysshe. *Shelley: Selected Poetry, Prose and Letters*. The Nonesuch Press, London. 1951.

Shepherd, Linda Jean. *Lifting the Veil*. Shambhala, Boston. 1993.

Sherrington, Sir Charles. *Man on His Nature*. Doubleday & Company, Inc., Garden City. 1955.

Siegel, Eli. *Damned Welcome*. Definition Press, New York. 1972.

Silcock, Arnold. *Verse and Worse*. Faber and Faber, London. 1952.

Simpson, George Gaylord. 'The Non-prevalence of Humanoids' in *Science*. Volume 143. 1964.

Simpson, George Gaylord. *Biology and Man*. Harcourt, Brace & World, Inc., New York. 1969.

Simpson, George Gaylord. *Concession to the Improbable*. Yale University Press, New Haven. 1978.

Simpson, George Gaylord. *Principles of Animal Taxonomy*. Columbia University Press, New York. 1961.

Simpson, George Gaylord. *This View of Life: The World of an Evolutionist*. Harcourt, Brace & World, Inc., New York. 1964.

Simpson, George Gaylord and Beck, William S. *Life: An Introduction to Biology*. Harcourt, Brace & World, Inc., New York. 1965.

Singer, Charles. *A Short History of Scientific Ideas to 1900*. Clarendon Press, Oxford. 1959.

Sinnott, Edmund W. *Cell and Psyche*. The University of North Carolina Press, Chapel Hill. 1950.

Skinner, B.F. 'A Case History in Scientific Method' in *The American Psychologist*. Volume 11. 1956.

Skinner, Cornelia Otis. *The Ape in Me*. Houghton Mifflin Company, Boston. 1959.

Slosson, Edwin E. *Keeping Up with Science*. Jonathan Cape, London. 1924.

Smiles, Samuel. *Thrift*. J. Murray, London. 1886.

Smith, Bertha Wilcox. 'Anonymous' in *Nature Magazine*. Volume 50, Number 5. May 1957.

Smith, J. Percy. *Selected Correspondence of Bernard Shaw: Bernard Shaw to H.G. Wells*. University of Toronto Press, Toronto. 1995.

Smith, William Jay. *Mr. Smith and Other Nonsense*. Delcorte Press, New York. 1968.

Smythe, Daniel. 'Small Flyers' in *Nature Magazine*. Volume 50, Number 6. 1957.

Snyder, Gary. *The Practice of the Wild*. North Point Press, San Francisco. 1990.

Snyder, Gary. *The Real Work*. A New Direction Book, New York. 1980.

Sontag, Susan. *On Photography*. Farrar, Straus and Giroux, New York. 1977.

Spencer, Herbert. *Social Statics*. John Chapman, London. 1851.

Spencer, Herbert. *The Principles of Biology*. Volume I. D. Appleton and Company, New York. 1886.

Spenser, Edmund. *The Complete Poetical Works of Edmund Spenser*. Houghton Mifflin Company, Boston. 1908.

Sperber, Michael A. and Jarvik, Lissy F. *Psychiatry and Genetics*. Basic Books, Inc., Publishers, New York.

Spinoza, Benedict de. 'Ethics' in *Great Books of the Western World*. Volume 31. Encyclopaedia Britannica, Inc., Chicago. 1952.

Squire, J.C. *Collected Poems*. Greenwood Press, Westport, CT. 1981.

Standen, Anthony. *Science is a Sacred Cow*. E.P. Dutton and Company, Inc., New York. 1950.

Stedman, E.C. *The Poetical Works of Edmund Clarence Stedman*. Houghton Mifflin and Company, Boston. 1884.

Steinbeck, John. *The Log From The Sea of Cortez*. Bantam Books, New York. 1971.

Steinbeck, John. *Travels with Charley*. The Viking Press, New York. 1962.

Stephens, James. *Collected Poems*. The Macmillan Company, New York. 1928.

Sterling, John. *Poems*. Edward Moxon, London. 1839.

Sterne, Laurence. *A Sentimental Journey*. J.M. Dent & Sons Ltd, New York. 1960.

Stevens, Peter F. *The Development of Biological Systematics.* Columbia University Press, New York. 1994.

Stewart, Ian. *Nature's Numbers.* Basic Books, New York. 1995.

Stockbridge, Frank B. 'Creating Life in the Laboratory' in *Cosmopolitan.* May 1912.

Strauss, Maurice B. *Body Water in Man.* Little, Brown and Company, Boston. 1957.

Strindberg, August. *Plays.* Secker & Warburg, London. 1970.

Strong, T.B. *Lectures on the Method of Science.* Clarendon Press, Oxford. 1906.

Stuessy, Tod F. *Plant Taxonomy.* Columbia University Press, New York. 1990.

Sturtevant, A.H. 'Social Implications of the Genetics of Man' in *Science.* Volume 120. September 10, 1954.

Sullivan, J.W.N. *The Limitations of Science.* A Mentor Book, New York. 1956.

Swift, Jonathan. 'Gulliver's Travels' in *Great Books of the Western World.* Volume 36. Encyclopaedia Britannica, Inc., Chicago. 1952.

Swift, Jonathan. *On Poetry.* Printed in Dublin. 1734.

Swinburne, Algernon Charles. *The Complete Works of Algernon Charles Swinburne.* Volume I. William Heinemann Ltd, London. 1925.

Szent-Györgyi. *Bioenergetics.* Academic Press Inc., New York. 1957.

Szent-Györgyi. *The Living State.* Academic Press Inc., New York. 1972.

Tabb, John Banister. *The Poetry of Father Tabb.* Dodd, Mead & Company, New York. 1928.

Tansley, A.G. 'The Classification of Vegitation and the Concept of Development' in *Journal of Ecology.* Volume 8, Number 2. June 1920.

Tansley, A.G. *Practical Plant Ecology.* G. Allen & Unwin Ltd, London. 1923.

Tarr, Rodger L. and McClelland, Flemming. *The Collected Poems of Thomas and Jane Welsh Carlyle.* The Penkevill Publishing Company, Greenwood. 1986.

Tax, Sol and Callender, Charles. 'Evolution After Darwin' in *Issues in Evolution.* Volume III. The University of Chicago Press, Chicago. 1960.

Taylor, A.M. *Imagination and the Growth of Science.* Schocken Books, New York. 1970.

Taylor, Bayard. *The Poetical Works of Bayard Taylor.* Houghton, Mifflin and Company, Boston. 1882.

Taylor Family. *Original Poems for Infant Minds.* Robert Carter & Brothers, New York. 1859.

Taylor, Gordon Rattray. *The Biological Time Bomb.* The World Publishing Company, New York. 1968.

Teale, Edwin Way. *Circle of the Seasons.* Dodd, Mead & Company, New York. 1953.

Teale, Edwin Way. *Near Horizons: The Story of an Insect Garden*. Dodd, Mead & Company, New York. 1944.

Temple, Frederick. *The Present Relations of Science to Religion*. Pamphlet, Oxford. 1860.

Tennyson, Alfred. *The Complete Poetical Works of Tennyson*. The Riverside Press, Cambridge. 1898.

Terborgh, John. *Diversity and the Tropical Rain Forest*. Scientific American Library. New York. 1992.

Terry, Rose. *Poems*. Ticknor and Fields, Boston. 1861.

Thackery, W.M. *The Complete Poems of W.M. Thackeray*. Frederick A. Stokes Company, New York. 1900.

Thaxter, Celia. *The Poems of Celia Thaxter*. Houghton, Mifflin and Company, Boston. 1898.

Thiery, Paul Henri. *The System of Nature*. Volume I. Garland Publishing, Inc., New York. 1984.

Thimann, Kenneth V. *The Life of Bacteria*. The Macmillan Company, New York. 1955.

Thomas, Lewis. *The Medusa and the Snail*. The Viking Press, New York. 1979.

Thomas, Lewis. *Late Night Thoughts on Listening to Mahler's Ninth Symphony*. The Viking Press, New York. 1983.

Thomas, Lewis. *The Lives of a Cell: Notes of a Biology Watcher*. The Viking Press, New York. 1974.

Thompson, D'Arcy Wentworth. *On Growth and Form*. Cambridge University Press, Cambridge. 1942.

Thompson, P. *The Life of William Thomson*. Volume II. Macmillan and Co., Limited, London. 1910.

Thomson, J. Arthur. *Concerning Evolution*. Yale University Press, New Haven. 1925.

Thomson, J.A. *Introduction to Science*. Henry Holt & Co., New York. 1911.

Thomson, James. *The Seasons*. Printed by A. Strahan, London. 1799.

Thomson, W. 'Scientific Laboratories' in *Nature*. Volume 31, Number 801. March 5, 1885.

Thoreau, Henry David. 'Walking' in *The Atlantic Monthly*. Volume 9, Number 56. June 1862.

Thoreau, Henry David. *The Writings of Henry David Thoreau*. Volume I. Houghton Mifflin Company, Boston. 1884.

Thoreau, Henry David. *The Writings of Henry David Thoreau*. Volume II. Houghton Mifflin Company, Boston. 1884.

Thoreau, Henry David. *The Writings of Henry David Thoreau*. Volume IV. Houghton Mifflin Company, Boston. 1884.

Thoreau, Henry David. *The Writings of Henry David Thoreau*. Volume V. Houghton Mifflin Company, Boston. 1884.

Thoreau, Henry David. *The Writings of Henry David Thoreau.* Volume VI. Houghton Mifflin Company, Boston. 1884.

Tiley, N.A. *Discovering DNA.* Van Nostrand Reinhold Company, New York. 1983.

Tille, Alexander. *The Works of Friedrich Nietzsche.* Volume XI. The Macmillan Company, New York. 1908.

Tiselius, Arne and Nilsson, Sam. *The Place of Value in a World of Facts.* John Wiley & Sons, Inc., New York. 1970.

Toffler, Alvin. *Future Shock.* Bantam Books, New York. 1971.

Tracy, Henry Chester. *American Naturalist.* E.P. Dutton & Co., Inc., New York. 1930.

Tribus, M. and McIrvine, E.C. 'Energy and Information' in *Scientific American.* Volume 225, Number 3. 1971.

Trotter, Wilfred. 'The Commemoration of Great Men' in *British Medical Journal.* February 20, 1932.

Trotter, Wilfred. 'Observation and Experiment and Their Use in the Medical Science' in *British Medical Journal.* July 26, 1930.

Turgenev, Ivan. *Fathers and Sons.* Translated by Michael R. Katz. W.W. Norton & Company, New York. 1994.

Turgenev, Ivan. *On the Eve.* Translated by Constance Garnett. The Macmillan Company, New York. 1920.

Twain, Mark. *A Connecticut Yankee in King Arthur's Court.* Grosset & Dunlap, Publishers, New York. 1945.

Twain, Mark. *A Double Barrelled Detective Story.* Harper & Brothers, New York. 1902.

Twain, Mark. *A Tramp Abroad.* Volume I. Harper & Brothers Publishers, New York. 1899.

Twain, Mark. *Following the Equator.* Volume I. Harper and Brothers Publishers, New York. 1899.

Twain, Mark. *Following the Equator.* Volume II. Harper and Brothers Publishers, New York. 1899.

Twain, Mark. *Mark Twain in Eruption.* Harper & Brothers Publishers, New York. 1940.

Twain, Mark. *Personnal Recollections of Joan of Arc.* Harper & Brothers Publishers, New York. 1896.

Twain, Mark. *The Complete Works of Mark Twain.* Harper & Brothers, New York. 1923.

Twain, Mark. *The Innocents Abroad.* The Library of America, New York. 1984.

Twain, Mark. *The Tragedy of Pudd'nhead Wilson.* Oxford University Press, New York. 1996.

Twain, Mark. *Roughing It.* Volume II. Harper & Brothers Publishers, New York. 1913.

Twain, Mark. *What is Man?* DeVinne Press, New York. 1906.

Tyson, Brian. *Bernard Shaw's Book Reviews.* Volume I. The Pennsylvania State University Press, University Park. 1991.

Urey, H. 'On the Early Chemical History of the Earth and the Origin of Life' in *Proceedings of the National Academy of Science.* Volume 38. 1952.

Vidal, Gore. *Armageddon? Essays 1983–1987.* Random House, New York. 1988.

Vogel, Steven. *Life in Moving Fluids.* Princeton University Press, Princeton. 1983.

Von Baeyer, H.C. 'Rainbows, Whirlpools, and Clouds' in *The Sciences.* Volume 24, Number 4. July/August 1984.

Waddington, C.H. *The Strategy of the Genes.* George Allen & Unwin Ltd, London. 1957.

Waddington, C.H. *Towards a Theoretical Biology.* Volume 2. Aldine Publishing Company, Chicago.

Walker, John. *Lectures on Geology.* The University of Chicago Press, Chicago. 1966.

Wallin, Ivan E. *Symbioticism and The Origin of Species.* The Williams & Wilkins Company, Baltimore. 1927.

Walters, Mark Jerome. *The Dance of Life.* Arbor House, New York. 1988.

Walton, Izaak. *The Compleat Angler.* Nathaniel Cooke, Milford House, Strand. 1854.

Ward, Barbara. *Who Speaks for Earth?* W.W. Norton & Company, Inc., New York. 1973.

Ward, Lester Frank. *Glimpses of the Cosmos.* Volume III. G.P. Putnam's Sons, New York. 1913.

Warner, Charles Dudley. *Backlog Studies.* Houghton, Mifflin and Company, Boston. 1892.

Watson, James D. 'Adaptation' in *Nature.* Volume 124, Number 3119. August 10, 1929.

Watson, James D. *Molecular Biology of the Gene.* Third edition. W.A. Benjamin, Inc., Menlo Park. 1976.

Watson, J. *The Double Helix.* Atheneum, New York. 1968.

Watson, James D., and Crick, Francis Harry Compton. 'Molecular Structure of Nucleic Acids' in *Nature.* Volume 171, Number 4356. April 25, 1953.

Webb, Mary. *The Spring of Joy.* Jonathan Cape, London. 1937.

Weintraub, Pamela. *The Omni Interviews.* Ticknor & Fields, New York. 1984.

Weinberg, Steven. 'Newtonianism, Reductionism and the Art of Congressional Testimony' in *Nature.* Volume 330, Number 6147. 3–9 December 1987.

Weiner, Jonathan. *The Next One Hundred Years*. Bantam Books, New York. 1990.

Weiss, P. *Hierarchically Organized Systems in Theory and Practice*. Hafner Publishing Company, New York. 1971.

Wells, Carolyn. *Baubles*. Dodd, Mean and Company, New York. 1917.

Wheeler, William Morton. *Social Life Among the Insect*. Harcourt, Brace and Company, New York. 1923.

Whitcomb, John C., Jr and Morris, Henry M. *The Genesis Flood*. The Presbyterian and Reformed Publishing Company. 1961.

White, Jonathan. *Talking on the Water*. Sierra Club Books, San Francisco. 1994.

White, T.H. *The Book of Beasts*. Jonathan Cape, London. 1954.

Whitehead, Alfred North. *Adventures of Ideas*. The Macmillan Company, New York. 1956.

Whitehead, Alfred North. *An Enquiry Concerning the Principles of Natural Knowledge*. Cambridge University Press, Cambridge. 1919.

Whitehead, Alfred North. *Process and Reality*. The Social Science Book Store, New York. 1941.

Whitehead, Alfred North. *Science and the Modern World*. The Macmillan Company, New York. 1929.

Whitehead, Alfred North. *The Organisation of Thought*. Williams and Norgate, London. 1917.

Whitman, Sarah Helen. *Poems*. Reeston and Round Co., Providence, RI. 1916.

Whitman, Walt. *Complete Poetry and Collected Prose*. Literary Classics of the United States, New York. 1982.

Whitman, Walt. *Specimen Days*. David R. Godine, Boston. 1971.

Whittaker, R.H. 'New Concepts of Kingdoms of Organisms' in *Science*. Volume 163. 10 January 1969.

Wilbur, Richard. *Things of This World*. Harcourt, Brace and Company, Inc., New York. 1956.

Wilde, Oscar. *De Profundis*. Division Books, New York. 1964.

Wilde, Oscar. *Intentions*. James Osgood McIlware & Co., 1891.

Wilde, Oscar. *Poems*. Roberts Brothers, Boston. 1882.

Wilde, Oscar. *The Complete Works of Oscar Wilde*. Harper and Row, New York. 1989.

Wilde, Oscar. *The Picture of Dorian Gray*. Charles Carrington, Paris. 1908.

Wilford, John Noble. *The Riddle of the Dinosaur*. Alfred A. Knopf, New York. 1985.

Willis, Nathaniel Parker. *Poems of Nathaniel Parker Willis*. George Routledge & Sons, Lim., London. 1891.

Willstätter, Richard. *From My Life*. W.A. Benjamin, Inc., New York. 1965.

Wilson, E.O. 'The Coming Pluralization of Biology and the Stewardship of Systematics' in *BioScience*. Volume 39, Number 4. April 1989.

Wilson, Edward O. *Biophilia*. Harvard University Press, Cambridge. 1984.

Wilson, Edward O. *Consilience: The Unity of Knowledge*. Alfred A. Knopf, New York. 1998.

Wilson, Edward O. *In Search of Nature*. Island Press, Washington, D.C. 1996.

Wilson, Edward O. *Sociobiology: The New Synthesis*. Harvard University Press, Cambridge. 1975.

Wilson, Leonard G. *Sir Charles Lyell's Scientific Journals on the Species Question*. Yale University Press, New Haven. 1970.

Wolpert, Lewis. *The Unnatural Nature of Science*. Harvard University Press, Cambridge. 1993.

Wood, J.G. *Common Objects of the Microscope*. George Routledge and Sons, Limited, London. 1861.

Wood, Robert Williams. *How to Tell the Birds from the Flowers and other Wood-cuts*. Dover Publications, Inc., New York. 1959.

Woodger, J.H. *Biological Principles: A Critical Study*. Kegan Paul, Trench, Trubner & Co., Ltd, London. 1929.

Woolley, Tyler A. *Acarology*. John Wiley & Sons, New York. 1988.

Wordsworth, William. *The Complete Poetical Works of William Wordsworth*. Houghton Mifflin Company, Boston. 1904.

Worster, D. 'The Ecology of Chaos and Harmony' in *Environmental History Review*. Volume 14. 1990.

Wright, Robert D. 'Letters' in *BioScience*. Volume 31, Number 11. December 1981.

Wright, Robert. *The Moral Animal*. Pantheon Books, New York. 1994.

Wynne, Annette. *All Through the Year*. J.B. Lippincott Company, Philadelphia. 1932.

Yogananda, Paramahansa. *Autobiography of a Yogi*. Self-Realization Fellowship, Publishers, Los Angeles. 1969.

Young, Michael. *The Metronomic Society*. Harvard University Press, Cambridge. 1988.

Young, Roland. *Not for Children*. Doubleday. Doran & Company, Inc., Garden City. 1930.

Zetterberg, J. Peter. *Evolution versus Creationism: The Public Education Controversy*. Oryx Press, Phoenix. 1983.

Zihlman, Adrienne L. 'Sex, Sexes, and Sexism in Human Origins' in *Yearbook of Physical Anthropology*. Volume 30. 1987.

Zimmerman, E.C. 'Distribution and Origin of Some Eastern Oceanic Insects' in *American Naturalist*. Volume LXXVI, Number 764. 1942.

PERMISSIONS

Grateful acknowledgment is made to the following for their kind permission to reprint copyright material. Every effort has been made to trace copyright ownership but if, inadvertently, any mistake or omission has occurred, full apologies are herewith tendered.

Full reference to authors and the titles of their works are given under the appropriate quotations.

A STUDY IN REALISM by John Laird. Copyright 1920. Reprinted by permission of the publisher, Cambridge University Press, Cambridge.

ACROLOGY MITES AND HUMAN WELFARE by Tyler A. Woolley. Copyright 1988. Reprinted by permission of the publisher, John Wiley & Sons, Inc., New York.

ALTERED FATES: GENE THERAPY AND THE RETOOLING OF HUMAN LIFE by Jeff Lyon and Peter Gorner. Copyright 1996, 1995. Reprinted by permission of the publisher, W.W. Norton & Company, Inc., New York.

AN ENQUIRY CONCERNING THE PRINCIPLE OF NATURAL KNOWLEDGE by Alfred North Whitehead. Copyright 1919. Reprinted by permission of the publisher, Cambridge University Press, Cambridge.

APES, ANGELS, AND VICTORIANS by W. Irvine. Copyright 1955. Reprinted by permission of the publisher, McGraw-Hill, New York.

ATOMS IN THE FAMILY by Laura Fermi. Copyright 1954. Reprinted by permission of the publisher, University of Chicago Press, Chicago.

BABOON MOTHERS AND INFANTS by Jeanne Altmann. Copyright 1980. Reprinted by permission of the publisher, Harvard University Press.

BEES, THEIR VISION, CHEMICAL SENSE, AND LANGUAGE by Karl von Frisch. Copyright 1950, 1971. Reprinted by permission of the publisher, Cornell University Press, Ithaca.

BEHAVIOR AND EVOLUTION by Anne Roe and George Gaylord Simpson. Copyright 1958. Reprinted by permission of the publisher, Yale University Press, New Haven.

BETWEEN PACIFIC TIDES by Edward F. Ricketts, Jack Calvin and Joel W. Hedgpeth. Copyright 1985. Reprinted by permission of the publisher, Stanford University Press, Stanford.

BIOPHILIA by Edward O. Wilson. Copyright 1984. Reprinted by permission of the publisher, Harvard University Press.

BOTANICAL WRITINGS by Johann Wolfgang von Goethe. Translated by Bertha Mueller. Copyright 1952. Reprinted by permission of the publisher, the University of Hawai'i Press, Honolulu.

COMPLETE POEMS by John Keats. Copyright 1982. Reprinted by permission of the publisher, Harvard University Press.

CONCESSION TO THE IMPROBABLE by George Gaylord Simpson. Copyright 1978. Reprinted by permission of the publisher, Yale University Press, New Haven.

CONDORCET: FROM NATURAL PHILOSOPHY TO SOCIAL MATHE-MATICS by Keith Michael Baker. Copyright 1975. Reprinted by permission of the publisher, University of Chicago Press, Chicago.

DARWIN AND THE MODERN SCIENCE by A.C. Seward. Copyright 1909. Reprinted by permission of the publisher, Cambridge University Press, Cambridge.

DESERTS ON THE MARCH by Paul B. Sears. Copyright 1935. Reprinted by permission of the publisher, University of Oklahoma Press, Norman.

DEVELOPMENT OF MICROBIOLOGY by Patrick J. Collard. Copyright 1976. Reprinted by permission of the publisher, Cambridge University Press, Cambridge.

DREAM SONGS by John Berryman. Copyright 1997. Reprinted by permission of the publisher, Farrar, Straus and Giroux, LLC.

ESSAY ON CLASSIFICATION by Louis Agassiz. Copyright 1962. Reprinted by permission of the publisher, The Belknap Press of Harvard University Press.

EVER SINCE DARWIN by Stephen Jay Gould. Copyright 1977. Reprinted by permission of the publisher, W.W. Norton & Company, Inc., New York.

SUBJECT BY AUTHOR INDEX

Amaranths such as crown the
maids..., 188

amaryllis
Tennyson, Alfred
A milky-bell'd amaryllis blew,
188

amoeba
Cudmore, L.L. Larison
Amoebas may not have
backbones..., 5
An amoeba is never torn apart
through indecision..., 5
Cuppy, Will
Amoebas not only divide, they
also blend, 5
Huxley, Julian
Amoeba has her picture in the
book..., 5
Popper, Karl
The difference between the
amoeba and Einstein..., 6
Unknown
An amoeba named Sam..., 6
When you were a soft
amoeba..., 6

analogy
Chargaff, Erwin
...it seeks refuge in allegory or
in analogy, 10
Emerson, Ralph Waldo
Science is nothing but the
finding of an analogy...,
10
Johnson-Laird, P.N.
...the discovery of a profound
analogy..., 10
Pepper, Stephen
The original area becomes his
basic analogy..., 10
Strindberg, August
...analogy, the highest form of
proof, 11

anemone
Bryant, William Cullen

...gray circles of anemones...,
188
Goodale, Elaine
O white anemone!, 189
Moore, Thomas
Anemones and seas of gold...,
189

animal
Ackerman, Diane
One of the things I like best
about animals in the wild
is..., 12
Agassiz, Louis
Animals are worthy of our
regard..., 12
Beston, Henry
We need another...more
mystical concept of
animals, 13
Borges, Jorge Luis
A certain Chinese
encyclopedia divides
animals into..., 13
Brophy, Brigid
I don't hold animals superior
or even equal to humans,
13
Bruchac, Joseph
Let my words be bright with
animals..., 13
Butler, Samuel
If one would watch them..., 14
Canetti, Elias
Whenever you observe an
animal closely..., 14
Ehrlich, Gretel
Animals give us their constant,
unjaded faces..., 14
Eiseley, Loren
Animals are molded by natural
forces..., 14
Eliot, George
Animals are such agreeable
friends..., 15

The true subject of science is
 the beauty of the world, 42

beaver

Outwater, Alice
 The beaver is utterly familiar,
  19

bee

Cleveland, John
 Nature's confectioner, the bee,
  248

Gay, John
 The careful insect 'midst his
  work I view..., 248

Purchas the Younger, Samuel
 Bees are political creatures...,
  248

Shakespeare, William
 So work the honey-bees..., 248

Smythe, Daniel
 The bees, those intergarden
  missiles..., 248

beech tree

Campbell, Thomas
 ...woodman, spare the
  beechen tree, 429

beetle

Crowson, Roy A.
 The beeetles are at once
  absolutely typical of...,
  249

Shakespeare, William
 And the poor beetle, that we
  tread upon..., 249

Wordsworth, William
 The beetle, panoplied in gems
  and gold..., 249

biologic categorization

Dunn, R.A.
 Biologic categorization is one
  of the most conspicuous
  aspects of..., 44

biological

Brower, David

A fallen tree supports a
 biological community...,
 44

Compton, Karl Taylor
 ...appropriate training for the
  handling of a great variety
  of biological situations...,
  44

Snyder, Gary
 ...the biological world is
  infinitely more complex,
  46

Trivers, Robert
 It's absurd not to use our best
  biological concept, 46

Wheeler, William Morton
 ...the actual, fundamental,
  biological structure..., 47

biological discovery

Unknown
 There is not the slightest
  chance of any biological
  discovery..., 46

biological lesson

Durant, Will
 ...the first biological lesson of
  history..., 45

biological life

Bird, J.M.
 We shall have a philosophy of
  biological life which..., 43

biological observation

Wesenberg-Lund, C.
 ...who first made a biological
  observation..., 46

biological research

Arber, Agnes
 ...the first step in biological
  research..., 43

biological rhythms

Young, Michael
 Every bodily process is pulsing
  to its own beat..., 48

Thomas, Lewis
   ...we derived...from some
      single cell..., 96
centipede
Blanshard, Brand
   The centipede was happy
      quite..., 20
chameleon
Wells, Carolyn
   The true chameleon is small...,
      381
champac
Moore, Thomas
   In her full lap the Champac's
      leaves of gold, 195
chance
Crick, Francis Harry Compton
   Chance is the only source of
      true novelty, 97
Darwin, Charles
   ...we are tempted to attribute
      their proportional
      numbers and kinds to
      what we call chance, 97
   ...working out to what we
      may call chance, 97
   I cannot think that the
      world...is the result of
      chance..., 97
du Nouy, Lecomte
   The laws of chance have
      rendered...immense
      services to science, 98
Keosian, J.
   ...the role of chance in the
      interaction of matter in
      the universe..., 98
LaPlace, Pierre Simon
   ...chance has not reality in
      itself..., 98
Monod, Jacques
   ...chance alone is at the source
      of every innovation..., 98
Reichenbach, Hans

...chance in combination with
      selection produces order,
      99
Thoreau, Henry David
   Who knows which way the
      wind will blow tomorrow,
      99
chaos
Adams, Henry Brooks
   Chaos was the law of
      nature..., 100
   ...chaos is all that science can
      logically assert..., 100
Blackie, John Stuart
   Chaos, Chaos, infinite
      wonder!, 100
Kant, Immanuel
   ...to fashion itself out of
      chaos..., 100
Santayana, George
   Chaos is perhaps at the bottom
      of everything..., 100
   Chaos is a name for any order
      that produces confusion
      in our minds, 100
Wilde, Oscar
   From eyeless Chaos..., 101
characteristics
Ardrey, Robert
   ...acquired characteristics
      cannot be inherited..., 102
cherry-tree
Longfellow, Henry Wadsworth
   In white with blossoming
      cherry-trees..., 429
chestnut-tree
Ingelow, Jean
   And when I see the chestnut
      letting..., 429
Lowell, Maria White
   The chestnuts, lavish of their
      long-hid gold..., 430
chigger
Hungerford, H.B.

millipede
Garstang, Walter
  The hatching of a Millipede
    brings curious things to
    light..., 28
mite
Duck, Stephen
  A crowd of dwarfish creatures
    lives..., 36
Frost, Robert
  But unmistakably a living
    mite..., 36
Hooke, Robert
  The least of Reptiles I have
    hitherto met with is a
    Mite, 36
Unknown
  The cheese-mites asked..., 36
moccasin flower
Goodale, Elaine
  The Indian's moccasin!, 206
mocking-bird
Longfellow, Henry Wadsworth
  ...the mocking-bird, wildest of
    singers..., 74
molecular biology
Chargaff, Erwin
  ...molecular biology is..., 295
Dobzhansky, Theodosius
  Molecular biology is Cartesian
    in its inspiration, 295
Kornberg, Arthur
  Molecular biology falters
    when..., 295
Luria, Salvador
  Molecular biology deals with
    questions of molecular
    structure..., 295
Maddox, John
  ...the unreflective state of
    molecular biology..., 295
Wolpert, Lewis
  The revolution in molecular
    biology..., 296

mollusc
Pallister, William
  Next, the MOLLUSCS present
    forty thousand kinds
    more..., 297
monkey
de Voto, Bernard
  Man is a noisome bacillus...,
    279
morning-glory
Jackson, Helen Hunt
  Generous in its bloom, and
    sheltering while it clings,
    Sturdy morning-glory, 206
Lowell, Maria White
  The morning-glory's
    blossoming..., 206
mosquito
Beaver, Wilfred
  Mosquitoes are like little
    children..., 256
Pallister, William
  The whole of Africa is our
    domain..., 256
moth
Carlyle, Thomas
  But see! a wandering
    Night-moth..., 256
museum
Belloc, Hilaire
  All in the Mu-se-um, 300
Edwards, R.Y.
  The physical heart of a
    museum is..., 300
Flower, Sir William Henry
  A museum is like a living
    organism..., 300
Goode, George Brown
  A finished museum is a dead
    museum..., 301
mutant
Crow, J.F.
  ...mutants would usually be
    detrimental, 302

Wormy apples at the
grocery..., 450
Isaiah 51:8
...and the worm shall eat them
like wool, 450
Pallister, William
Then the WORMS seven
thousand of species can
show..., 451
Taylor Family
No little worm, you need not
slip..., 451
Unknown
...if you drink alcohol, you
won't have worms, 451
...except the worms. They
came in apples, 452
wormwood
Thoreau, Henry David
...I perceive/The Roman
wormwood..., 215
wren
Wordsworth, William
...the little wren's/In snugness
may compare, 84
yak
Belloc, Hilaire
As a friend of the children
commend me the Yak...,
35
Smith, William Jay
The long-haired Yak..., 35
yew
Wordsworth, William
Of vast circumference..., 437
zoo
Esar, Evan
Another place where people
may visit..., 453
Hediger, Heini
Zoo biology is still a very

young science..., 453
...the most frequent
misconceptions which is
constantly met in the
zoo..., 453
Wynne, Annette
Excuse us, Animals in the
Zoo..., 453
zoological
Elton, Charles
...sending the whole
zoological world flocking
indoors..., 454
Queneau, Raymond
In the dog days while I was in
a bird cage..., 456
zoological chart
Feynman, Richard P.
You mean a zoological chart!,
455
zoologist
Bock, W.J.
Communication...among
zoologists is the core of
zoological
nomenclature..., 454
Esar, Evan
The only one who can tell the
difference between..., 455
Polanyi, Michael
The existence of animals was
not discovered by
zoologists..., 455
zoology
Bierce, Ambrose
Zoology is full of surprises, 454
The father of Zoology was
Aristotle..., 454
Kavaleski, V.O.
...the task of modern
zoology..., 455

AUTHOR BY SUBJECT INDEX

smelt, 184
termite, 257
toucan, 83
tsetse, 254
Needham, James G.
biology, 58
living, 275
Needham, Joseph
organization, 359, 360
Nelkin, Dorothy
creationism, 118
Newman, Joseph S.
chromosomes, 103
Darwinian, 122
hereditary, 232
man, 282
primordial ooze, 375
Newton, Sir Isaac (1642–1726)
English mathematician
nature, 337
Nicholson, Jack
God, 230
Nicholson, Norman (1914–?)
Writer/poet
fungi, 217
Nietzsche, Friedrich (1844–1900)
German philosopher
species, 402
Norse, Elliot A.
water, 443
Novacek, M.J.
systematist, 411

-O-
Oemler, Marie Conway
butterfly, 249
Oken, Lorenz
animal, 16
Oliver, Bernard M.
exobiology, 171
Oliver, Mary
nature, 337
Olson, S.L.

classification, 107
Oparin, A.I. (1894–1980)
Russian biochemist
extraterrestrial life, 171
Orgel, Irene
God, 230
Osborn, Henry Fairfield
conservation, 113
Osler, Sir William (1849–1919)
Canadian physician
biology, 58
Outwater, Alice
beaver, 19
Ovid
sea, 354

-P-
Page, Jake
name, 306
Pallister, William
bird, 64
crustacean, 119
euglena viridis, 24
extraterrestrial life, 171
fish, 180
humming bird, 72
insect, 245
molluscs, 297
mosquito, 256
photosynthesis, 367
protozoa, 378
spider–scorpion, 38
tadpole, 8
variation, 439
worm, 451
Pascal, Blaise (1623–1662)
French scientist
God, 230
man, 283
Patterson, John W.
creationism, 118
Pauling, Linus (1901–1994)
US chemist
mutation, 303